Praise for *Ecological Footprint*

How much of the Earth's biocapacity are humans using? Before the
Ecological Footprint, no one knew for sure. Now, we have the equivalent
of an instrument panel. This is a momentous development, and one
that all citizens of the planet should know about—because the warning
lights are flashing red.

— Richard Heinberg, Senior Fellow, Post Carbon Institute,
author of *The End of Growth* and *Peak Everything*

This excellent book helps us understand and truly appreciate Nature's
capacity to continue to provide life support for our planet and for all of
its inhabitants.

— Julia Marton-Lefèvre, Former Director General,
International Union of Conservation of Nature

The great challenge for humanity in the 21st century is not to stop the
growth of the economy, but to stop the growth in the human ecological
footprint—and eventually bring it back to one planet. This important
book by the inventors of the concept explains why and how.

— Prof. Jorgen Randers, former President of the BI Norwegian
Business School, coauthor of *The Limits to Growth*,
author of *2052—A Global Forecast for the Next Forty Years*

If we did to our bank account what we have been doing to the Earth's
natural capital we would have been bankrupt long ago. The planet has
been extremely lenient with us but that resilience is about to give way to
a natural and human crisis. This book is a loud wake up call to everyone.

— Christiana Figueres, former Executive Secretary, UNFCCC

In the nick of time, as humanity crashes up against the resource
limitations of our collective twenty first century lifestyles, *Ecological
Footprint* provides a clear eyed and accessible analysis of the challenge.
With clarity and compassion, *Ecological Footprint* reveals both our
alarming self-inflicted situation and the way forward. Wackernagel
and Beyers' well written book has the power to turn the tide.

— Thomas E. Lovejoy, Professor of Environmental Science and Policy,
Institute for a Sustainable Earth George Mason University

A superb treatment of Ecological Footprint accounting as a part of our global balance sheet. Regardless of whether you are a student, a teacher, or an economist, you will find much of substantial importance in this book.

— Peter H. Raven, President Emeritus, Missouri Botanical Garden, St. Louis

Looking for a science based practical tool to navigate your future on Earth? Here it is. *Ecological Footprint* provides an integrated and concrete measure of our human pressure on the Planet. We all urgently need to reconnect our lives to planet Earth and adopt a biocapacity approach to modern life, translating it into concrete steps of how each and every one of us can contribute to building resilient and sustainable societies.

— Prof. Johan Rockström, Director of the Potsdam Institute for Climate Impact Research

Human societies live—and die—by cultural myths including the catastrophic modern myth of perpetual economic growth. With their update of the *Ecological Footprint*, Wackernagel and Beyers wield the most effective myth-busting tool available. Failure to read this book should disqualify any would-be elected official from running for office.

— William Rees is a human ecologist, ecological economist and former Director of UBC's School of Community and Regional Planning in Vancouver. He co-developed the ecological footprint with Mathis Wackernagel

At a time when we must find ways to urgently respond to the existential threats of climate change and ecosystems extinction, this book systematically lays before us the accounting metric necessary to evaluate a world in overshoot. Almost 50 years ago two of the "Limits to Growth" scenarios predicted global system "overshoot and collapse" by the mid- to latter-part of the 21st century. Today ecological footprinting, if adopted by governments and business leaders, alongside comparable planetary solutions, could help navigate our emergence from emergency.

— Sandrine Dixson-Declève , Co-President, The Club of Rome

Ecological Footprint

Managing Our Biocapacity Budget

Mathis Wackernagel • Bert Beyers

GLOBAL FOOTPRINT NETWORK

new society
PUBLISHERS

Cover design by Diane McIntosh.
Cover photo (leaves) and illustration (footprint): ©iStock
Initial English Translation by Katharina Rout.
Interior illustrations by Phil Testemale.

Printed in Canada. First printing September 2019.

Inquiries regarding requests to reprint all or part of *Ecological Footprint* should be addressed to New Society Publishers at the address below. To order directly from the publishers, please call toll-free (North America) 1-800-567-6772, or order online at www.newsociety.com

Any other inquiries can be directed by mail to:

New Society Publishers
P.O. Box 189, Gabriola Island, BC V0R 1X0, Canada
(250) 247-9737

LIBRARY AND ARCHIVES CANADA CATALOGUING IN PUBLICATION

Title: Ecological footprint : managing our biocapacity budget /
by Mathis Wackernagel and Bert Beyers.

Other titles: Footprint. English

Names: Wackernagel, Mathis, 1962– author. | Beyers, Bert, author. |
Rout, Katharina, translator.

Description: Translation of: Footprint: Die Welt neu vermessen. | Translated by Katharina Rout. | Includes bibliographical references and index.

Identifiers: Canadiana (print) 20190099534 | Canadiana (ebook) 20190099542 |
ISBN 9780865719118 (softcover) | ISBN 9781550927047 (PDF) |
ISBN 9781771423007 (EPUB)

Subjects: LCSH: Environmental economics. | LCSH: Economic development—Environmental aspects. | LCSH: Human ecology. |
LCSH: Nature—Effect of human beings on. | LCSH: Environmental protection. |
LCSH: Conservation of natural resources. | LCSH: Sustainability.

Classification: LCC HC79.E5 W3313 2019 | DDC 333.7—dc23

Funded by the Government of Canada   Financé par le gouvernement du Canada   Canada

New Society Publishers' mission is to publish books that contribute in fundamental ways to building an ecologically sustainable and just society, and to do so with the least possible impact on the environment, in a manner that models this vision.

new society PUBLISHERS

Certified (B) Corporation

FSC www.fsc.org   MIX
Paper from responsible sources
FSC® C016245

Contents

Prelude

by Mathis Wackernagel

This book is not really about the Ecological Footprint. Rather it is about *biocapacity*—our planet's biological power to regenerate and reproduce plant matter. This primary productivity of nature is the source for all life, including human life.

Biocapacity is not an invention or a method, the same way gravity isn't. Both are a force of nature that we can observe and measure.[1]

The importance of biocapacity is rising. Given climate change and resource constraints, biocapacity, or rather how we manage it, is increasingly determining humanity's future. Humanity's poor stewardship of biocapacity has made it the materially most limiting factor for the human enterprise. Understanding biocapacity's relevance therefore empowers us to build countries, cities, or economies that can thrive for good, rather than being marred by surprises.

This is why I dedicate this book to the foresters, farmers, conservationists, park rangers, and fishery managers of the world, and in particular Fritz Jenni, an astute farmer from Langenbruck, Switzerland, who generously took me under his wings from my early boyhood to my teenage years, and taught me about nature's cycles, miracles, and powers. Fritz Jenni's love for the land and its animals, especially all people, continues to inspire me. He helped me see how biocapacity is the ultimate force enabling everything we do. Thank you, Fritz!

My dear friend Bert Beyers developed the German version of this book with me after spending a multi-month sabbatical with us in Oakland. This English version is updated and amended, and I hope you will enjoy it as much as Bert and I enjoyed the journey of making our ideas accessible to those who care about our planet's future.

I am also deeply grateful of all those who have accompanied me on the journey. One who stands out is Bill Rees, who evolved from being a teacher to being a friend. Working with him as we developed the first version of the Ecological Footprint during my PhD dissertation in the early 1990s was deeply enriching. Another indefatigable "partner in crime" is Susan Burns, without whom the Global Footprint Network venture would not have flourished. She started Global Footprint Network with me in 2003, shortly after giving life to our son André. Susan's vision, dedication, unstoppable energy, and relentless quest for what's next have made more possible than I could have imagined.

My wonderful colleagues in the proper and extended Global Footprint Network have put strong legs on our initially rather wobbly ideas. They ended up contributing even better ideas. The community starts with our fabulous advisors and board members who dedicate their time and energy as noble volunteers to the mission, many researchers who came to us as interns and left as friends, and the amazing staff members who have put in so much more than what we ever could give them back. Uncountable partners have brought intriguing projects to life and participated in events and campaigns, touching hundreds of millions, if not more. Supporters, donors, and funders have given to our cause generously and selflessly, which moves me particularly. They could have used those resources for anything else, including increasing their own comforts. But they chose to bet on an idea about how we can all thrive within the means of our one Earth. Such dedication gives me hope that humanity has the capacity to build a far better future.

But the ultimate dedication goes to planet Earth, my one and only.

FOOTPRINT

Why?

What good is a plane without cockpit dials? Sure, it flies. But how high, how fast, and in which direction? What is its exact position? In rough weather or at night, flying without instruments becomes sketchy. Even in good weather. Particularly if basic instruments—like the fuel gauge—are missing. Without knowing how much fuel is left in the tanks, any flight becomes unsafe.

Operating an economy is similar. Like a plane, an economy is fueled. The difference is that an economy not only requires kerosene but also devours coal, food, timber, water, and many other materials our planet provides. How many resources does it take for each breakfast, vacation, or new apartment each person may enjoy? How much nature does a city, a power plant, a nation, or the entire human enterprise use? If we are so utterly dependent on all these resources, how come our economies do not have fuel gauges?

In our daily lives, we pretty much know the dollar value of everything. Why? Because our financial budgets are limited. We want to know what we can afford. Like our own financial budget, nature's resource budget is limited too. And the mother of all resources, the most limiting resource, is, as we will see, the biological assets—our planet's *biocapacity*. So how much nature can we afford? And if indeed nature's budget is limited, why don't we measure it?

Would you get on a plane that does not have a fuel gauge? If not, how come we continue operating countries without having an equivalent gauge? How resource secure is your country? And what is the trend?

One possible answer is that we did not have a reasonable instrument for measuring our demand on nature. Also, for a long time, no tool was needed, since nature appeared to be immense and endless. It is different today. Now nature's limits have become obvious, whether it is groundwater depletion, climate change, or decline of the oceans' fish stocks.

Good measurement instruments finally exist: with the Ecological Footprint, we can measure our use of nature. It offers a basic ecological accounting system. While for the economy, money is typically used as the accounting unit, the Footprint uses *biologically productive surfaces of the Earth* as its currency. These surfaces harbor the most significant resource on our planet: the capacity of Earth to renew itself. On surface areas, photosynthesis transforms sunlight, water, nutrients into plant matter, over and over again. Therefore, every demand of the economy on nature's ability to produce and renew plant matter can be expressed as the corresponding surface area needed to meet this demand. Yield figures tell us how much cropland, a forest, or grazing land provides each year. This is the demand side of the story.

We can also measure with ever greater precision what nature supplies, thanks to modern technology. Satellites deliver us up-to-date images of our planet. They show where forests, cities, streets, deserts, lakes, pastures, or grasslands are located. Those satellite pictures can be verified by direct measures in the field. On-the-ground measures track, for instance, how many potatoes or how much wheat is actually grown. At the country level, United Nations statistics provide detailed numbers for most of these resource flows: land areas, yields of the various land types, amounts produced and traded, population size, energy use, and so on.

Financial accounts always look at two sides: income against expenditures, or assets against liabilities. Footprint accounting is tracking demand on nature against what nature renews. It is a basic, straightforward, science-based description: How much nature is available (income)? How much nature do people use (expenditure)?

To manage the ecological capital of our planet merely on gut instincts does not make much sense. No one would bring their money to a bank that does no bookkeeping. A bank statement gives us an objective financial review—a status report. This is exactly what's needed for the resource situation of the planet at this time. That's the reason why the Footprint primarily aims at government and business decision-makers. But these accounts also need to be understood by the citizenry so they can hold their decision-makers accountable.

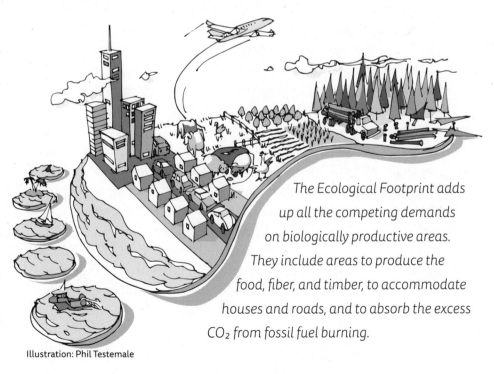

The Ecological Footprint adds up all the competing demands on biologically productive areas. They include areas to produce the food, fiber, and timber, to accommodate houses and roads, and to absorb the excess CO_2 from fossil fuel burning.

Illustration: Phil Testemale

The Footprint reveals how much of our planet's productive area is used for each human activity. Complex processes can be summed up in one single number, similarly as money gets reported in simple numbers like Return on Investment or Revenue versus Costs. This boils complex issues down to their essence and makes them accessible. It allows us to negotiate. The Footprint, therefore, is not only a communication tool that is intuitively understood by a broad public. It also serves as a transparent tracking tool to measure the performance of policies and the implications of decisions in public and private domains.

The parallels between economy and ecology goes beyond their names. In both domains, mismanagement is characterized by spending more than you earn. Physics and value creation have to go hand in hand: how can the value of real estate expand continuously, even if the actual real estate object does not change? How can we continuously accumulate a huge amount of debt and hope that somehow sometime it can be paid off? How can we continuously increase money supply without adding commensurate tangible value (even Google searches are "material," and so are bitcoins, digital photographs, or iTunes). How can we presume that expansion works forever? How can we expect an economy to forever deliver more, without expanding the natural capital needed to fuel the economy accordingly? How come we commonly forget that income generation depends on resource availability?

According to the latest Footprint calculations, humanity overused nature's biological budget (the biocapacity of the planet) by 75% in 2019. In other words, humanity uses nature currently 75% faster than it renews. This overuse is called ecological *overshoot*. Most estimates predict that the global population will rise from about 7.7 billion today to 9 or 10 billion in 2050. And the residents of the BRICS countries (Brazil, Russia, India, China, South Africa) will continue to work hard to raise their standards of living. And so will many others, in spite of potential economic turmoil. All these forces turn resource security into the central challenge of the 21st century.

Some might wonder whether we are in the midst of water, climate, fisheries, or food crises. The answer is that all those crises root in the same cause: our tremendous hunger for resources. This becomes evident as we examine the human resource metabolism a bit closer.

With rising ecological pressures, everybody, from individual to company, city or country, has "skin in the game" since no one can operate without the availability of sufficient resources. While demand is still going up, a lot of ecosystems on the planet are already overused and weakened. Such overuse destabilizes climate, guts fish stocks, or erodes biological productivity. It threatens adequate access to food and water for all. It might lead to conflicts, migration, economic hardship.

The financial crisis of 2008 provided the planet's ecosystems with a little breather. Resource and waste flows did not grow as fast as before. In some areas they are even declining. But such forced decline is not the goal. Because overshoot will end. Peter Victor astutely observes that humanity will only benefit if we end overshoot "by design, not disaster."[1] In other words: How can we decelerate humanity's metabolism without strangling the economy? How do we strengthen our resource security without burdening those who are struggling economically and are left behind? How fast can we end overshoot while ensuring high quality of life for all? How slowly can we end overshoot without putting everybody's well-being at risk?

The answer is simple. Ecological health and human well-being is not a real trade-off. Rather, resource security is the enabler of lasting human progress. Yet often we are tempted to believe that sustainability is just about keeping everything as it was. In high-income cities of Europe, for instance, many are under the false impression that maintaining things as they are is a workable strategy. So many architectural details of Paris and London look exactly the same as they did 100 years ago. This continuity covers up the rapid change characterizing the world and leaves inhabitants of those places in an illusion. In reality, the world is shifting at the speed of light: Consider, for instance that just during the lifetime of the authors, humanity has burnt 80% and 84% respectively of all fossil fuels ever used. What portion of fossil fuel ever used was burnt during your life?

Year you were born	Percentage of fossil energy burned since then	Year you were born	Percentage of fossil energy burned since then	Year you were born	Percentage of fossil energy burned since then
1896–1905	96	1972	73	1996	43
1906–12	95	1973	72	1997	42
1913–18	94	1974	71	1998	40
1919–23	93	1975	70	1999	39
1924–28	92	1976	68	2000	37
1929–33	91	1977	67	2001	36
1934–37	90	1978	66	2002	34
1938–41	89	1979	65	2003	32
1942–45	88	1980	64	2004	31
1946–48	87	1981	63	2005	29
1949–51	86	1982	62	2006	27
1952–54	85	1983	61	2007	25
1955	84	1984	59	2008	23
1956	84	1985	58	2009	21
1957–58	83	1986	57	2010	19
1959–60	82	1987	56	2011	17
1961	81	1988	54	2012	15
1962–63	80	1989	53	2013	13
1964	79	1990	52	2014	11
1965–66	78	1991	50	2015	9
1967	77	1992	49	2016	7
1968	76	1993	48	2017	4
1969	75	1994	46	2018	2
1970–71	74	1995	45	2019	0

Figure I.1. What percentage of all the fossil energy ever used throughout human history has been burned since you were born? Here a scandalous fact about Justin Bieber: during his short life, 46% of all fossil fuel ever used was burnt (he was born in 1994). The figures are for 2019. If you read the table in 2020, then go back one year as a first estimate, two years, if you read this two years after 2019. This means, for Justin Bieber the approximate result will be 46% in 2020, 48% in 2021, and so on.

During Mathis's life, the global population has more than doubled, the pressure on nature tripled. History is playing itself out at breathtaking speed. That turns one question into our central challenge: How can all thrive within the means of our one planet?

The Footprint metric delivers some navigational support.

For example, Footprints capture cities' use of nature for all it takes to make the city tick: food, housing, heat, light, mobility, and waste management. If the Footprint of a resident of a compact Mediterranean city like Siena or Salamanca is only ½ or ⅓ of a resident living in the sprawling city of Canberra, Atlanta, or Los Angeles, then Siena undoubtedly has an admirable and significant advantage. Those who prepare themselves better for a world with resource constraints (or already live in cities with such built-in advantage) will have a much better chance to thrive. Those who hesitate to adapt will struggle. Implementing a thoughtful long-term resource policy is in your own self-interest, whether you are a city, a region, or a nation. It is needed right now. Los Angeles will not become a Siena overnight.

Since most of humanity lives in cities (which also concentrate CO_2 emissions and consumption),[2] it is cities that will largely define the fate of human civilization in this century. As we shape our cities with every infrastructure update, housing project, traffic policy, the Footprint can help to make investment choices fit for the future. Let's look at traffic, for example: As complex as the discussion about buses, trains, and cars, connection and steering of the systems may be, the Footprint reduces all this information to one single number: the required biologically productive area to fuel these systems. That's something one can work with. The Footprint thus isn't just a measurement but also a management tool.

Human cities and communities need to ask themselves: Where do we get our energy from? Our food? How much do we use compared to our competitors? How much do we use compared to that available per capita in the world? A reoccurring topic is efficiency: Are we already using all our possibilities to live better with fewer resources?

For regions and countries, the supply side (their biocapacity) as well as their resource management is at least as critical. What is our resource base? How much biocapacity is in our territory? Using more biocapacity than available within the boundaries of our country pushes us into biocapacity deficit spending. In contrast, having more biocapacity available than we use leaves that country with an

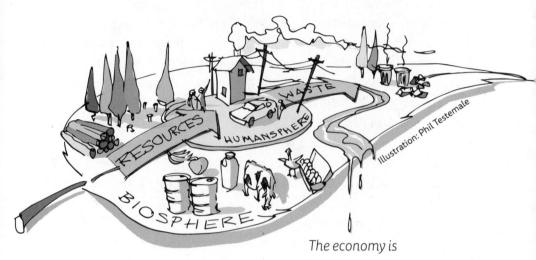

Illustration: Phil Testemale

The economy is
embedded within nature.
It is a "wholly owned subsidiary" of the biosphere.
All material ingredients come from Earth, and all used
up materials that are not recycled are returned to Earth.
Therefore, Earth's regenerative capacity is the materially most
limiting factor for the human enterprise.

ecological reserve. The more countries and regions know about their biocapacity balance, the stronger their ability to guide and accommodate the radical change that will become part of our landscape. There is no doubt: The tightening competition for our planet's biocapacity will be a major challenge in the future.

> *The message of the Footprint is: We can measure not only the availability of nature, but also human demand on it. Knowing both sides gives us the full picture of our ecological foundation and empowers us to manage our destiny. It's a practical tool for those wanting to prevent ecological bankruptcy in the 21st century.*

The Footprint is a descriptive indicator. It can monitor the course of events and show whether the chosen path is producing the hoped-for success or not. Footprint numbers are also free of moral preconceptions or imposed values. They don't tell anybody what they should or shouldn't do. They empower us merely to consider how much biocapacity is available, how much we use and who uses what, and what the implications might be for us and others. At the end of the day, Footprint analysts are motivated by the idea that thriving lives for all are possible within the means of our one planet. Pooran Desai from Bioregional call this "one planet living."[3]

Ultimately, we cannot escape the fact that humanity, with all its activities and in all areas of life, is part of nature. It is a dependency we cannot break. Yes, there are some philosophical and religious scriptures that try to tell us otherwise: That humans are separate from nature, that nature can and must be subdued, that it needs to be exploited and "civilized." As a result, humanity has subjugated and dominated ever more of nature, pushed back pristine parts to an extent that overuse has systematically become the norm. We have maneuvered ourselves into a cul-de-sac. Even evangelical ministers are now making the case that preserving creation is true worship.[4]

The Footprint is an accounting system that documents our ecological performance, nothing more and nothing less. By revealing the limits of nature, it contributes to building a globally sustainable economy. Its science-based description of what we use and what we have will hopefully feed into a consensus on where to go. By making visible basic physical boundary conditions, it helps define the playing field for societies and economies. Sustainability will only become reality if economic incentives are aligned with ecological possibilities. At the moment, they obviously are not.

Today, the most comprehensive Footprint accounts track the performance of countries. They cover all countries for which complete (or near complete) data sets in the United Nations statistics exist.[5] Over 245 national entities exist, and many are small. The 190 largest house about 99% of humanity. To assess its Footprint and biocapacity, each country is tracked using up to 15,000 data points per year. For 194 national entities, there are enough data in the UN statistics to produce the Footprint results.

The accounting method is not alarmist. On the contrary, it intentionally errs on the side of underreporting overshoot. Humanity's demand—its Ecological Footprint—is undercounted as not all demands are captured in UN statistics. The biocapacity side, however, is most likely overestimated since some damaging activities such as soil erosion or groundwater loss are not yet factored into the current accounts for lack of comprehensive and consistent data. This means that, in reality, the biocapacity deficits are most likely larger than what current accounts report, as discussed in more detail later in the book.

The Footprint is a highly aggregated measure of people's use of nature. The accounts capture a broad array of aspects. The Footprint captures all the aspects that compete for the productive surfaces of this planet. This leads to the method's communicative power: it summarizes human demand in one single number. It also compares overall demand to overall availability. It views everything from the perspective of the planet's ability to regenerate, its biocapacity. For instance, the use of fossil fuels like coal, gas, and oil is also included in terms of the biologically productive area such as the forests needed to absorb the resulting carbon dioxide emissions. Absorption of those greenhouse gases is one of the competing demands on the planet's biocapacity. There are trade-offs. Simply explained: we could either absorb more carbon dioxide or grow more carrots.

The Footprint is not the only ecological indicator. It does not claim absolute coverage, nor a monopoly. It rather focuses on one specific, yet fundamental question: How much of the biologically productive capacity of the planet is being used to power the human enterprise? For other relevant questions, other methods are needed. Like the different navigational instruments in the cockpit of an airplane, which are complementary to each other. Ultimately, we need a few clear and robust metrics that are simple to understand and can be used by many. We need a common "currency" that bundles the complexity of human dependence on nature, and makes choices comparable. This is the Footprint's ambition.

The Footprint framework affirms human vitality and aspiration: People want to live, and live well. But to thrive depends on how humanity will manage its ecological home. The challenges are considerable. They require us to employ all creativity and ingenuity we can get. In this context, the Footprint is a decisive tool to provide foresight, and to unleash our intellectual and innovative power.

Who Is behind This Effort

Mathis Wackernagel and William Rees conceived the Footprint in the early 1990s at the University of British Columbia. The method has evolved and been deployed widely by hundreds of cities, over a dozen countries, uncountable institutions as well as international agencies, such as the European Commission, the European Environment Agency, the International Organisation of La Francophonie, and the United Nations, including its Convention on Biological Diversity.

Global Footprint Network, with headquarters in Oakland (California), was established in 2003 to steward the methodology, develop standards, advance the accounts, and find novel applications in collaboration with partner organizations around the world. In 2018, Global Footprint Network joined forces with York University in Toronto (Canada) to build a global academic network that will host and maintain the National Footprint and Biocapacity Accounts as an independent, public-benefit venture, thereby improving accessibility, independence, and robustness of the results.[6]

In 2005, Global Footprint Network set itself the goal to have at least ten national governments officially test the Footprint before 2015.[7] In 2012, the Philippines and Indonesia became country number ten and eleven. More than twelve national government agencies have tested the Footprint already. They mostly concluded that the accounts adequately reflect their reality.[8]

There is still a long way to go. Switzerland, for instance, held a referendum on the Ecological Footprint in September 2016 where its citizens were asked whether Switzerland should strive to reach a Footprint by 2050 that could be replicated worldwide (currently it would take over three Earths if everybody on the planet lived like the Swiss).[9]

A breakthrough would be to have United Nations agencies adopt the Footprint broadly and contribute to its improvement, standardization, distribution, and application. Imagine if the world community realized that we need an instrument to measure our physical dependence on our planet, in the ways that GDP measures economic activity.

Our dependence on nature needs to be measured, in physical units. This is not unprecedented. Not everything in public policy is measured in financial units. For instance, we don't measure unemployment, longevity, or population size in dollars either.

There are significant options for shaping our various pathways. Even more so if we have access to the most relevant information and ideas. And possibly most importantly, the courage and wisdom to implement them. Here's the good news: The Footprint doesn't make our life more difficult—it enables us to make our cities and countries livable and our successes long-lasting. If you accept physics, knowing the biocapacity of one's own country and managing its Footprint will be as important in the future as financial accounting already is today. Knowing biocapacity is beneficial, as it is helpful to know about gravity. Knowing gravity still does not make it easier walking uphill. But it helps us build more robust houses and stronger bridges. Knowing about biocapacity and having robust Footprint accounts works for us. It gives us foresight and enables us to build a future that serves us all.

PART I

FOOTPRINT

The Tool

CHAPTER 1

AREA AS CURRENCY

How Much Biocapacity Does a Person Need?

Everyone, big or small, has an Ecological Footprint. How much nature people need depends on what they eat, how they dress, what their home is like, how they move around, and how they get rid of their waste. All of that can be measured. The resulting data allows us to determine the area of biologically productive land and water that is required to grow food, produce fiber for clothing, build houses to shelter people, and absorb their waste. We can measure the carbon dioxide from burning coal, gas, and oil. In the end, we all live on what the "global farm" provides, and we can accurately measure what the farm provides, and what people consume.

Everyone understands money. People with money have more options, and possibly fewer worries, at least material ones. Those with enough money can live how and where they like. Everyone welcomes them. As long as they can pay, no one will show them the door. We can do many things with money. For example, we can compare things. Money also tells us how much everything costs. Once we know the prices, we can relate them to our income. How long do I have to work so I can afford this mobile phone? How much do I earn, compared to my expenses? Compared to last year? Or compared to the income of someone in Singapore?

Ecological Footprint accounting is a tool that, like money, asks the core question: How much nature does everything cost? How much

How the Footprint Works:
Just Think of a Farm

The productive area of a farm is the farm's biocapacity. What it can produce is determined by the area, as well as the productivity of each acre. In the US, pastures are sometimes measured in "cow-calf acres"—how many cow-calf pairs can be maintained on one acre. It is both the area and its productivity that counts.

The Ecological Footprint estimates how much farm it takes to produce what we consume, including everything we eat, all the fiber and timber we use, all the space to house our roads and buildings, and to absorb all our CO_2 waste from burning fossil fuel. There is competition for our farm's productive areas as a farmer can't graze cows where she places her house, and can't plant tomatoes where she builds her pond.

A farm family may want to know how hungry they are for food, materials, heating fuel compared to what the farm can provide. We can create the same comparisons to the world, countries, regions, cities, and even individuals.

Humanity's biggest farm is our planet. Thanks to Ecological Footprint accounting, we come to realize that the way we operate our "farm" now is out of balance, as our collective demand exceeds by at least 70% what our planet's ecosystems replenish.

Nature can make up for the difference by depleting stocks. Examples are cutting timber faster than it regrows, emitting more CO_2 than the planet's ecosystems absorb, pumping up more groundwater than is being recharged, or catching more fish than restocks. This business model only works so long—whether for farmers or humanity as a whole.

As you look at the world from a biological perspective,
you start to recognize that every country is essentially
a farm with forests, pastures, cropland, etc. How big is
this farm compared to the resource demand of its residents?

Illustration: Phil Testemale

biocapacity is required for a glass of orange juice, and how much for a liter of gas? And we can go further: How much nature does a person need? A person's Footprint is a "currency" which is spent to provide services, to offer space for our buildings, to produce goods and to dispose of them. For a person, their Footprint is the sum total of all they require, including their waste (because waste too draws on nature). What the Euro, Dollar, or Yuan is to money, the hectare—or more precisely the global hectare—is to the Ecological Footprint.[1]

Just as different currencies can be set off against each other, so can the Footprint's area units. This is the point: that there is a single unit—a *tertium comparationis*—that everything refers to. Obviously, not every global hectare is identical, only sufficiently similar. But the same is true for money since one dollar for a person with minimum wage means something quite different than one dollar for a billionaire.

Therefore, in the same way one financial figure cannot describe the health of an economic entity, mapping the entire ecological reality with just one number is obviously crude and insufficient. In fact, Ecological Footprint accounting is not suggesting it is mapping the entire ecological reality. Rather it puts emphasis on biological resources (as we will discuss in more detail). The reason is that *biological resources* are materially more limiting for the human enterprise than the non-renewable resources like oil or minerals. For instance, while the amount of fossil fuel still underground is limited, even more limiting is the biosphere's ability to cope with the CO_2 emitted when burning it. The burning and coping are competing uses of the planet's biocapacity. Similarly, minerals are limited by the energy available to extract them from underground and concentrate them.

Figure 1.1. Ecological Footprint in global hectares per person, by country, 2016 data. In 2016, the world's biocapacity averaged 1.63 global hectares per person. Credit: Global Footprint Network—National Footprint and Biocapacity Accounts 2019 edition, data.footprintnetwork.org.

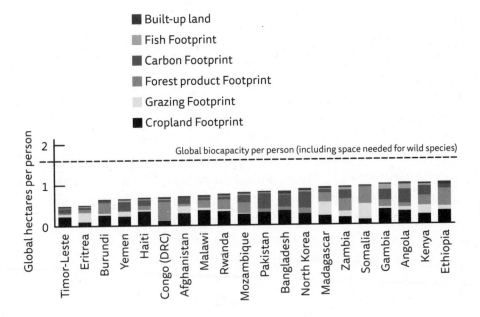

Since biology and area are interconnected, Ecological Footprint accounts take areas of biologically productive land or water as their measurement unit. As we will see, such a simple unit makes communication more accessible, and our situations more understandable. Prices allow people to communicate with others about the high or low cost of a good. The Footprint enables us to have productive dialogues about the different ways we consume nature: about high or low consumption, about its impact on this or that ecosystem—summarized as one single number, the sum of all our demands on nature.

Let's visit a department store. Just as the goods on offer carry price tags that identify their monetary value, and just as food products come with information about nutrients and ingredients, all products could come with an additional number that identifies the biocapacity that has gone into the product. The front of the price tag would tell us what we must pay, while the back would tell us how much nature was used. A block of cheese, a pair of jeans, a holiday trip—everything can be measured in biocapacity: what size of area is required to provide this product or service? For cheese, it is mainly the grazing land a cow needs to produce milk and of course the energy needed to turn milk into cheese. For jeans, it is the cotton field. Trips are enabled by many things, from aviation or car fuel to electricity for the trains, food, maintenance and cleaning of the hotel and washing of the linen. For many city dwellers, electricity may seem to come magically from the socket and milk from a carton, but behind everything we use there is a piece of nature.

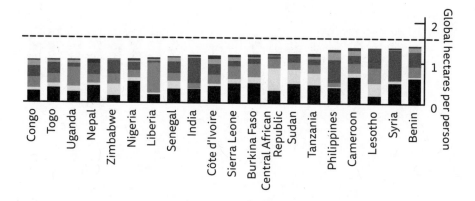

Here, too, we have a parallel to money: as long as we have enough, all seems well and we take it as a given. But what if there isn't enough? To have no biological capacity feels not that different from having no money. If, for example, you are stranded in a foreign city without cash or credit card, what will you eat? Where will you sleep?

What would happen if nature all of a sudden could no longer provide its wonderful services? If there wasn't enough water to support life and economic activity in the first place? What if the oceans' fishing grounds shrank or even collapsed while demand for fish continued to rise and fish became rarer and more expensive? What if the fields in one's backyard couldn't produce enough to sustain one's family and people—like many in rural Bangladesh—didn't have the money to buy additional food? What if the forests and oceans one day all of a sudden no longer absorbed carbon dioxide but instead released the gas they had stored into the atmosphere? What then?

Money is our core economic measure for assessing value. But money can do more than simply measure value: it is also a means of payment and, as such, gets passed from person to person. The Footprint can't do that. We can exchange the fruits of biocapacity, for example by importing timber and exporting meat. People or trade

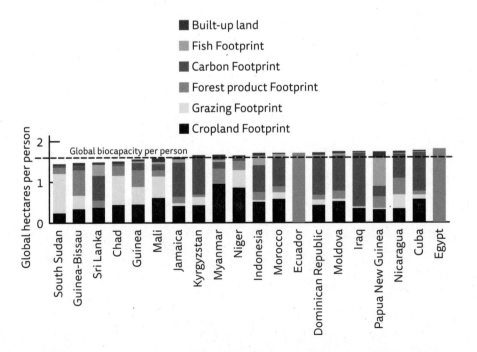

statistics may not recognize that, since it is not actual Footprints units that get traded. Rather, we can measure the Footprints of timber and meat that is traded.

Money is also a kind of storage system for one's assets (as in a savings account or a portfolio), but that, too, is different with the Footprint. Nature's assets always exist in nature itself, and the Footprint, as an accounting method or a code number, only measures and identifies them. Whereas money is recognized if not idolized as valuable, nature's capital is undervalued. We behave as if nature were infinite and inexhaustible in its provision of riches to humanity. In the long term, however, it is nature that is the most valuable asset, whereas money is just a symbol.

Of course, things exist that we cannot buy, such as true love. We cannot assign it a monetary value. Another example is the atmosphere. People have developed the habit of treating our atmosphere as a free garbage dump for their emissions. As with money, there are areas where the Footprint does not apply. A rock, for example, has no Footprint. It simply is, and its existence requires no measurable consumption. Animals, on the other hand, do have a Footprint; they breathe, drink, and feed, consume biocapacity and hence area. A fish eaten by a seal is no longer available to us, or only indirectly when in turn we eat the seal or use its pelt.

How much biocapacity do we need? In order to eat, to clothe ourselves, to build our houses and heat them, also to travel and to transport any goods, we need the supplies that nature provides. In the

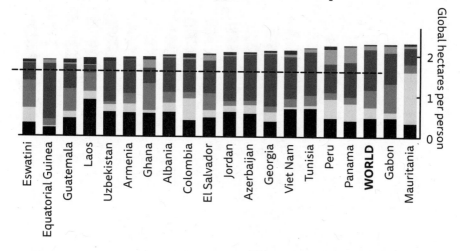

process, we leave behind solid, liquid, and gaseous waste. Nature has to cope with that too. As we move through the world, we leave behind our "Footprint." Some of us tread with heavy steps, while others have such a small and light step they hardly touch the ground. But every human being, big or small, leaves a trace as long as they live. It is this trace that the Footprint metaphor refers to.

The Ecological Footprint measures not only the demands an individual puts on nature but can equally be applied to the population of cities, nations, or humanity as a whole.

Let's take fossil energy as an example: Since the Industrial Revolution, we have availed ourselves of massive amounts of nature's resources of coal, oil, and gas, when in fact these are non-renewable resources, or to be more precise, resources that renew themselves only over enormous periods of time. We extract them from the Earth's crust and bring them to the surface and hence into the biosphere. For Footprint calculations, the amount of coal or oil underground does not enter into the equation. After all, these materials are not part of living nature but came to be over millions of years; in that sense, they

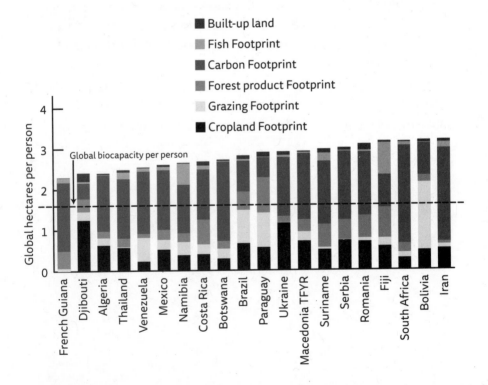

are assets more like a piece of gold or a painting by Picasso. Also, they turn out to be rather plentiful compared to what the biosphere can handle. It is by using coal or oil that we consume nature, and this consumption is what Footprints measure.[2] When such quantities of fossil energy are burned, carbon dioxide is released. And then our biosphere has to cope with that, because this is new carbon dioxide that previously was not part of the natural cycles.

To prevent an increasing concentration of carbon dioxide in the atmosphere that will lead to a long-term destabilization of our climate, that additional carbon dioxide should be removed—but, so far, only a small portion has been removed. The remainder we leave for nature to cope with it. A good percentage of the excess carbon dioxide is now being absorbed by the oceans (and further acidifies them), some is absorbed by ecosystems on land, but some land uses also lead to net emissions. A lot of the carbon dioxide is left in the atmosphere and accumulates. The Footprint method therefore asks: how big an area, how much forest, is necessary to absorb the remaining amount of carbon dioxide? Research shows that an average hectare of forest on this planet, if managed for climate protection, can annually absorb roughly the same amount of carbon dioxide as is released by burning 900 liters (or 240 gallons) of gasoline.[3]

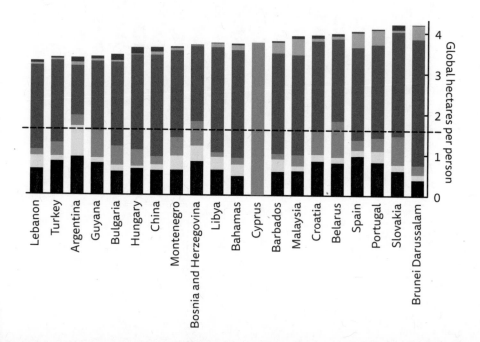

Over the past 200 years, the atmosphere's carbon dioxide level has risen by about ⅓ from 278 ppm to more than 410 ppm, and more if we include other greenhouse gases. We are obviously not dedicating enough of the planet's biological capacity—mainly forests and oceans—to sequester the combustion residues as quickly as we generate them.[4] One reason is that there are many other competing demands for the planet's biological capacity as well. Plus, there isn't enough to do that: Recently, the carbon Footprint has become so large that it alone is now exceeding the Earth's regenerative capacity.

Still, if we deploy area to sequester more carbon dioxide, we could have considerably less biocapacity left for other purposes, such as the production of food, fiber, or fuelwood, or the creation of urban areas. Grazing and crop agriculture can in some ways sequester car-

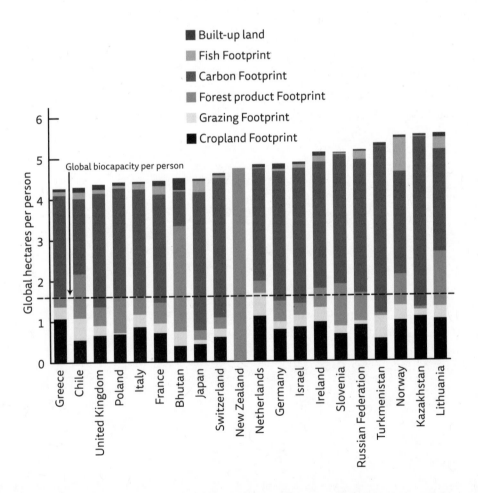

bon. We can point to promising experimentation, but the experiments have not yet scaled.[5] Maintaining yields while also sequestering carbon dioxide would show up as a very welcome increase in biocapacity.

The situation is similar with other non-regenerative resources such as steel, copper, or minerals. These materials are indirectly connected with the living part of nature; we extract most mineral substances from the Earth's crust. Extracting, concentrating, and processing them puts demands on living resources. Since the Footprint accounts for impact on living resources, metals and minerals are included in terms of the biocapacity it takes to mine them, and the energy used for extraction, transport, and processing. This

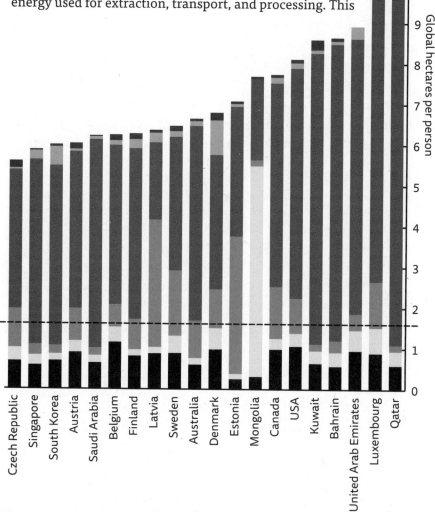

is the demand metals and minerals put on nature. And that, in turn, brings us back to carbon dioxide and to the biocapacity required to store carbon through photosynthesis in solid biomass. Put differently, mineral substances and ores are valuable assets like gold or shares, but in contrast they consume additional energy to make them available to the human economy. That energy, too, requires biocapacity.

For a long time, most people paid attention primarily to the non-renewable resource aspect of natural capital. People recognized that the supply of fossil energy sources as well as of ores and minerals is ultimately finite, that they will sooner or later be exhausted, or that certain resources are left only in low concentrations, making it too hard to extract them. This concern is understandable, given that industrial production processes depend on such materials. Indeed, some of these materials have already become rare. But recently we have come to realize that renewable resources with their life-supporting functions are even more scarce, and that even though they can be replenished, they can also be depleted.[6]

Renewable resources—forests, fish stocks, wetland—can be entirely used up through overuse. This happens eventually when people exploit renewable resources faster than they can regenerate.

Figure 1.2. Biocapacity in global hectares per person, by country, 2016 data. In 2016, the world's biocapacity averaged 1.63 global hectares per person. Credit: Global Footprint Network—National Footprint and Biocapacity Accounts 2019 edition, data.footprintnetwork.org.

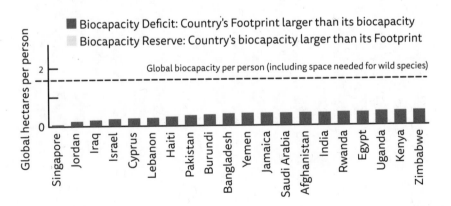

Whereas most non-renewable resources are less fundamental for the support and conservation of life, renewable resources are a *conditio sine qua non*, a non-negotiable condition, for the existence of all life on Earth. For this reason, it is especially the renewable resources, and with them the biosphere's overall potential to regenerate, that constitute the materially most limiting factor for human life and well-being. This constraint is shared, of course, with the world's more than ten, or possibly hundred, million animal and plant species.

In short, the Footprint looks at the world as if it were a farm. How big is it? How much does it yield? How much do we use, compared to what the farm grows? Farmers, too, take area as their point of reference—and it is those very same areas that provide the ecological services upon which life depends.

A farmer's perspective on nature can be translated into a science-based accounting system. The framework behind Ecological Footprint accounting brings together millions of numbers culled from satellites, trade statistics, censuses, and questionnaires. The United Nations has created comprehensive data sets for the entire world that has tracked the world consistently since 1961. The UN stamp makes the data official, and turns them into the most neutral and accepted data set for comparing nations. This data set makes possible to calculate with consistency the Footprint of nations all the way back to 1961. Today, Global Footprint Network calculates them for all the 220 countries included in the UN statistics. Out of the data sets, about 194 are complete enough to produce results, at least for one year. For every single country, and for every year, the method presently requires up

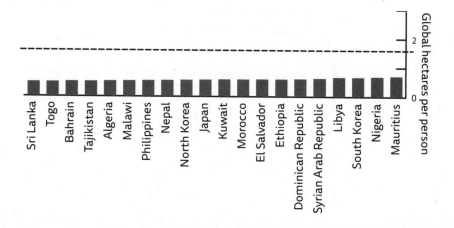

to 15,000 bits of data stemming from all kinds of data sets: energy, agricultural production, land use, population, fisheries, forest, and so on.[7]

This way, every country gets a number that indicates its residents' average consumption of nature's resources—their Footprint—as well as an estimate of the ability of the country's natural environment to renew what people demand—its *biocapacity*.

To repeat the question: How much biocapacity does a person occupy? Today we can answer this question with ever better statistics, even though we know for sure that our answers remain somewhat imprecise because reality is just too detailed, and even the best statistics cannot capture everything. Even though they are not absolutely exact, still our answers point in the right direction, can be verified and improved upon. They are merely the best available answers to our questions. Because Global Footprint Network, its partner organizations, and other institutions continue to improve the science involved, the results are increasingly reliable, too.

This is also the reason why Global Footprint Network, together with York University in Toronto, is now gathering a coalition of countries, supported by a rigorous global academic network, that will own and produce the future editions of independent, transparent, and robust National Footprint and Biocapacity Accounts.[8] This will be more powerful than just having one organization, Global Footprint Network, produce the accounts. Then, the results will be more trusted and seen as unbiased, which makes it more likely they will inform public and private decision-making.

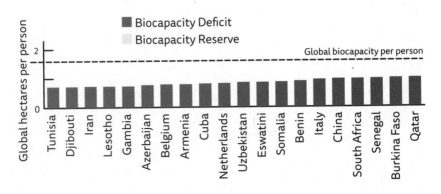

The results for the 2019 edition, the latest ones at the time of printing, cover all the countries up to 2016. (The time lag reflects the time delay in UN data compilation.) The Footprint captures each person's "global farm"—forests, fishing grounds, grazing lands, croplands—that this person needs for his or her resource consumption, waste absorption, and to accommodate the buildings and roads she or he occupies. They show that the average Footprint of a person in Haiti—a country whose ecological devastation and intense setback through an earthquake was accompanied by economic turbulence and intense political upheaval—is 0.68 global hectares. The demand for biocapacity in Kenya or Uganda amounts to 1.0 and 1.2 global hectares per person, respectively. A German, on the other hand, claims on average 5.0, a Frenchman 4.7, an American 8.3, and a resident of the United Arab Emirates 10.2 global hectares.

A number of Footprint calculators are available online that let individual people easily calculate their own Footprint. Not to brag, but we like ours best.[9] We like it not just because it uses cute graphics, but because it is directly calibrated against national calculations. Also, it builds on easy questions anybody can answer without having to get up and look at utility bills or weigh their garbage for a week. Like any quiz, it asks simple questions about your nutrition—for example, how many times a week you eat meat—about features of your house, and about your mobility habits. The answers allow for a rough estimate of your individual Footprint, including the translation of how many Earths it would take if everybody lived like you. It even tells you the date of Earth Overshoot Day, if all people on our planet lived like you.

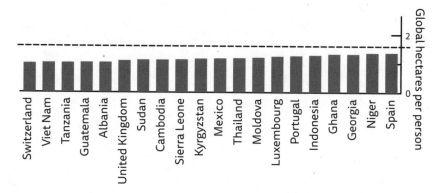

Calculating the Ecological Footprint of Cardi B in Six Easy Steps

Let's take singer Cardi B to illustrate how Footprints are calculated. Say Cardi B's coffee comes from Guatemala, the wheat to feed the chickens that lay her eggs comes from Iowa, and the wool used for her jacket is from New Zealand. Thus her Footprint is spread all across the world.

To assess her current Footprint, we track:

1. How much pasture does it take to feed the cows for the dairy and meat she consumes this year, the wool she uses, and the leather for her shoes, jackets, and furniture?
2. How large are the fields needed to produce all her beans, cotton, rubber, sugar, cereals—not only for her croissants and spaghetti, but also for feeding the chickens and pigs she might eat this year? How much for the cotton and silk?
3. How much ocean area is necessary to produce the fish that she eats this year?
4. How much land for her homes (or portion of them, if she shares her homes with others), her gardens, and her share of the roads, city squares, airports, and parks?
5. How much forest area is necessary to absorb the CO_2 from fossil fuel she uses this year—for heating and cooling her homes,

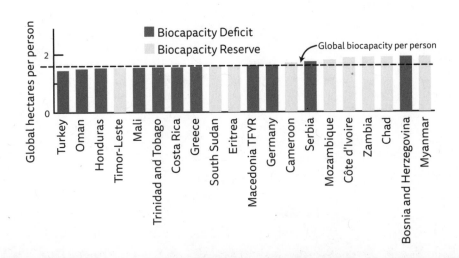

producing the goods and services she consumes, driving and flying her around?

6. How much area is needed for the energy and resources used to provide Cardi B's share of social expenditures like hospitals, police forces, government services, educational facilities and museums, and military activities?

To get Cardi B's Footprint, we first itemize all the areas from the above questions—all the actual areas needed for everything she uses. Then, we translate every actual area into standardized *global hectares* with world average productivity or growing potential. Hectares that are highly productive, let's say three times more than world average, would be counted in this case as three global hectares. These global hectares become the common currency that allows us to compare all hectares on an equal footing. Then we simply add all those global hectares up and get her Footprint for this year. Voilà.

This is the area Cardi B occupies for the entire year to provide what she consumed in the entire year. Next year, her Footprint will be different again as her consumption, technological efficiency, as well as the productivity of the biosphere may change. Check out your own at footprintcalculator.org.

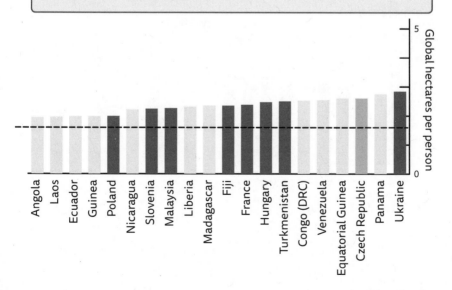

> The method, however, is applicable not only to lifestyles but also to any other activity, product, and service, from a shower, a piece of bread, to a breakfast or an air trip or a doctor's visit.

The Footprint method gives us a new perspective. We can now see the actual physical "costs" of the things we use day in and day out. Some of those things give us a rich and fulfilled life. Others we just use out of habit. For each thing, we can see how much biocapacity it requires. It shows in numbers how our individual existence is directly linked to the planet's ecological capacity, something city dwellers sometimes forget. With this fresh perspective, we realize that the flows of materials and energy are not somewhere out there, separate from the economic realm. Instead, the numbers show how these resources flow through our lives. It makes obvious how human life and our economy are subsystems of the biosphere. The Ecological Footprint method is a tool that details the physical metabolism of humans and nature; it is both a micro- and a macro-instrument. On both a small and a large scale, we can quantify what nature provides and how we consume its provisions.

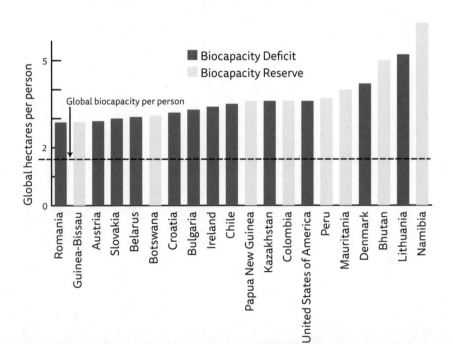

Footprint accounts mainly describe what is. The fact that we can even measure biocapacity and therefore make it tangible and specific is key. For example, Footprint accounting makes obvious that there are competing uses for the planet's limited biocapacity. It puts the challenges of climate change in context with many other demands. Sequestration of emitted carbon dioxide and other greenhouse gases competes with producing food or fiber. And if we do not have enough capacity for that,

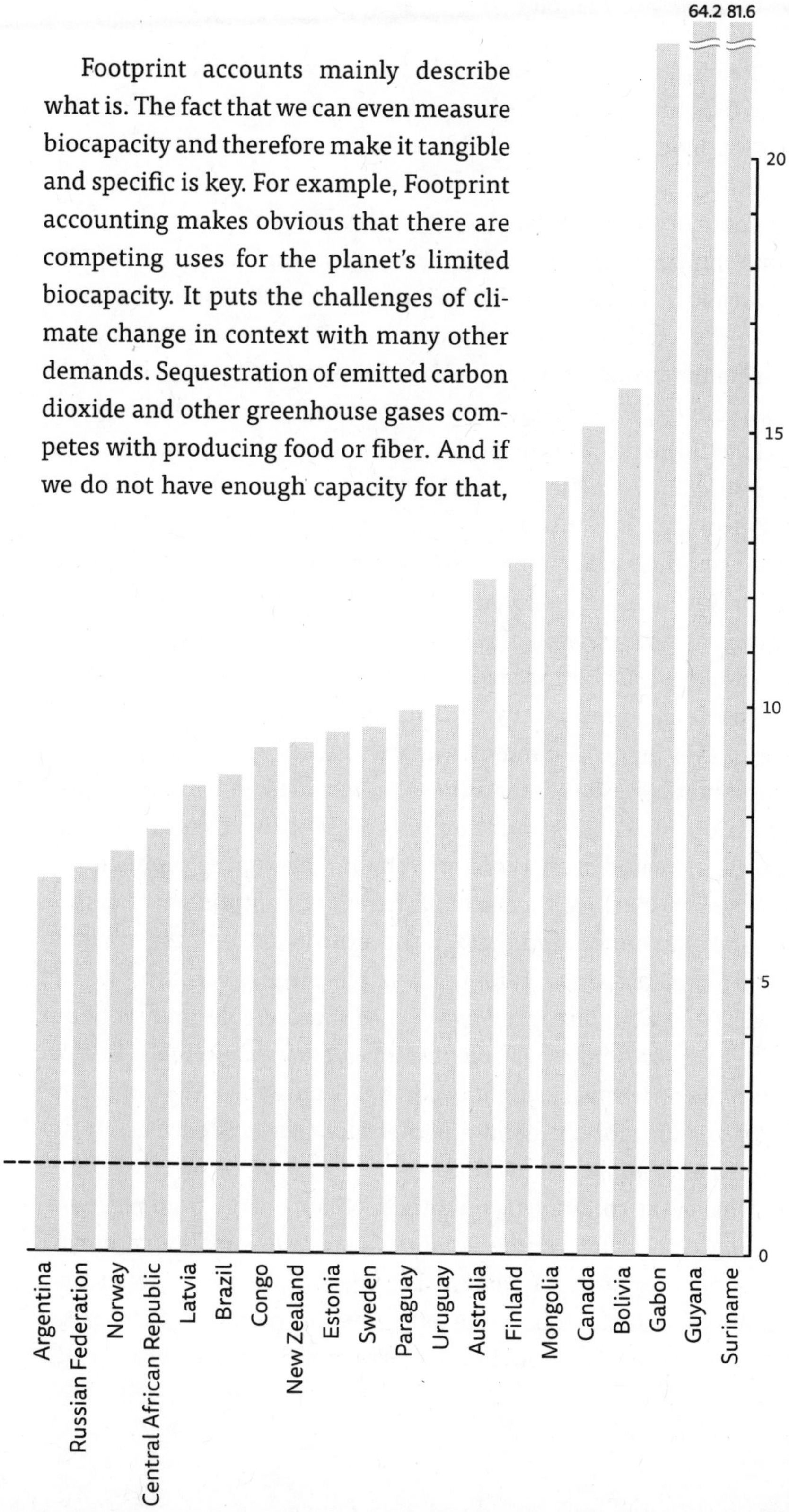

64.2 81.6

20

15

10

5

0

Argentina
Russian Federation
Norway
Central African Republic
Latvia
Brazil
Congo
New Zealand
Estonia
Sweden
Paraguay
Uruguay
Australia
Finland
Mongolia
Canada
Bolivia
Gabon
Guyana
Suriname

Earth's biocapacity deteriorates. For instance, carbon dioxide builds up in the atmosphere, which over time could significantly erode the planet's biocapacity through changing and more volatile weather patterns.

About 60% of humanity's Footprint today is the result of our consumption of fossil energy. Our carbon Footprint has grown rapidly. About 150 years ago, at the onset of the coal and steam revolution, humanity's carbon Footprint was essentially zero. Since 1961, when reliable data collection by the United Nations started, it has more than doubled. Our energy consumption has grown even faster, especially for natural gas, which emits less CO_2 when combusted and therefore has a smaller carbon Footprint per energy unit than coal or petroleum. But this climate benefit only holds true, though, if little of natural gas' methane escapes uncombusted. Methane itself is a powerful greenhouse gas. Even small methane losses in the extraction and distribution of gas—in the order of 2%—void the climate advantage of gas compared to coal.[10]

The demand for resources hardly knows an upper limit. We can live in ever bigger homes, own more residences, and drive cars or ride planes almost as much as we like, provided we have the money. As for our food, transportation of food products over greater distances, increased meat consumption, and ever more sophisticated food preparation are increasing humanity's Footprint, too. Carbon dioxide emissions accumulate in the atmosphere and contribute to lasting climate change. With the help of Footprint accounts, we can assess what would happen if we drew a considerable amount of our energy supply from renewable resources, such as agrofuels. In most cases, the atmosphere might become less burdened—but would we perhaps shift more demand on other biological systems? Footprint assessments would capture that.

Most of the common techniques to obtain energy from regenerative sources—water, wind, and biomass energy—emit less carbon dioxide, yet they often also require biologically productive areas, even windmills that stand on cropland. Different methods exist for the extraction of energy from biomass. For biofuels, typically the fruits of agricultural produce are used, such as corn grains, canola, rapeseed or palm oil kernels. Second-generation methods, on the other hand, use the entire plant with a correspondingly higher efficiency while ideally not competing with food crops. However, they have not become viable yet. The Footprint can quantify the demand on nature per unit of energy for each method.

The climate implications also become more obvious from a Footprint perspective. Fossil fuel allowed humanity to overcome biocapacity constraints; cheap fuels became versatile alternatives to the products of the planet's biocapacity. Fossil fuel is not just high-quality energy but can be used to produce plastics, fibers, and chemical products. The cheap and plentiful fossil energy also enabled the intensification of agriculture. In the US, it takes currently about 6 calories of fossil fuel to produce 1 calorie of food.[11]

In return, the emission from fossil fuel use, particularly CO_2, has overwhelmed the capacity of the biosphere to absorb this gas. The result is greenhouse gas accumulation in the atmosphere. If we allow for further accumulation of carbon in the atmosphere, the possibility of climate instability increases. This could erode food production, since agriculture depends on predictable climate. It was the magically stable climate conditions of the 10,000 years of the Holocene that enabled the emergence of agriculture. Therefore, the prospects for humanity are brighter if we stop fossil fuel use very soon and learn how to live only off biocapacity. Eventually we have to live only from what the planet can regenerate, so the more effectively we prevent climate change, the more biocapacity we will have.

These interactions between climate, fossil fuel, and biocapacity reveal the challenge of global warming and the significance of biocapacity. These interactions emphasize the reality of our astoundingly robust yet vulnerable planet Earth. We are biological beings on a biological planet. To bring all these aspects together is at the heart of the Footprint.

The Footprint functions like a map. It provides a description of the physical reality in which we live. As defined by its core principle, it translates human demands on our ecosystems into a common denominator. An extensive data set exists in the background, just as with a map. But the map shows only the essentials: cities, roads, borders. If it were to show every single tree or house, we could no longer read it. This reduction of complexity of Footprint accounting allows us to capture an intricate and convoluted reality. Like a map, the Footprint enables us to better understand and navigate in our world with its complex and diverse life-support systems. It helps us evaluate risks and opportunities; it supports us in finding a viable path forward.

CHAPTER 2

ECOLOGICAL HINTERLAND

How Much Biocapacity
Does a City Need?

For the first time in human history, more people live in urban than in rural areas. Although the settled areas make up only a small part of the planet's surface, it is already obvious: Earth's future will be determined primarily in cities. All depends on how cities provide their residents with water, food, and energy, and how their architecture, settlement, and infrastructure are shaped. How do cities value the natural capital they depend on? In times of dwindling resources, a larger question emerges: What does it take for all to flourish within our planet's resource budget? Footprint accounting points us in the right direction.

In a remote area of Arizona, a gigantic dome-shaped structure was erected in 1987–89. Inside was a closed ecosystem, with savannahs, tropical rain forests, a mangrove swamp, areas of water to represent the oceans, but also with intensely cultivated croplands and residential buildings. The project's name, Biosphere II, deliberately pointed at the living environment that makes up our planet, Biosphere I. The purpose of this experimental structure was to create a self-sustaining biotope and to gather experience so we would know, for example, what one day crewed bases on the Moon or Mars might require.

On September 26, 1991, the airlock closed behind the first eight inhabitants of Biosphere II. For about two years, they lived in futuristic-looking buildings that had underground connections with each other. They were all hermetically sealed off from the outside world. As time went by, life in the artificial world grew increasingly difficult. It turned out that the project's reinforced concrete, as it was still hardening, was using up the artificial atmosphere's oxygen and was emitting carbon dioxide. Eventually, oxygen had to be added from the outside. More surprises followed. Microbes in the soil raised the atmospheric levels of nitrogen and carbon dioxide in Biosphere II more than expected. Cockroaches and spiders particularly enjoyed the high-tech buildings and rapidly multiplied. Obviously, it is not that easy to imitate, start up, and maintain nature's complex regulatory systems.[1]

Let's do a mental experiment and superpose the image of Biosphere II onto any modern city, such as Shanghai, Berlin, Dubai, or New York.[2] A gigantic glass bowl, turned upside down, forms a dome over the city. Neither air nor water nor food, neither energy sources such as oil or gas nor building materials such as bricks or sand can enter the artificial biotope from outside. It is hermetically sealed. Even wastewater, car exhaust, and household garbage remain trapped under the glass dome. Only sunshine has unimpeded access to the futuristic city, so at least there's light during daytime. With the sunlight, a certain amount of energy permeates the dome. Whether insects or rodents experience this artificial city as a paradise—nobody knows.

This mental experiment approximates the idea of the Ecological Footprint. The decisive question is: how big would the glass dome above the city actually have to be in order to supply the residents with everything they need to survive? Or put more simply, how much biocapacity does a city need?

In fact, we now have sufficient scientific knowledge for a fairly exact answer to this question. London has been footprinted multiple times. One of the first "City Limits," published in 2002, still has its antiquish website.[3] The study, at that time, concluded that the average resident of London required 6.6 global hectares—or roughly eight soccer fields—of biologically productive area to maintain their habitual level of consumption, including being housed, moved, and governed.[4] The garbage produced every year by the private households, industry, and construction projects of the British capital would fill the enormous Royal Albert Hall (with a height of more than forty meters) 265 times.

It is also noteworthy how much the city's residents eat every single day: that study found that their food amounted to a hefty 41% of London's total Footprint. If one calculates the sum total of the area required by all of London's residents, one comes up with 49 million global hectares, which is 300 times the geographical expanse of the city and more than half the biocapacity of the entire United Kingdom.

Of course, the resources a city avails itself of are both local and global in nature, their origins scattered across the planet. This has long been true for London. During the mid-19th century, the capital of the British Empire already had four million residents, more than any other city in the world. Even then it had a Footprint unrivalled in history. Stanley Jevons, a prominent British economist of the 19th century described it as follows: "The plains of North America and Russia are our corn-fields; Chicago and Odessa our granaries; Canada and the Baltic are our timber-forests; Australasia contains our sheep-farms, and in South America are our herds of oxen; Peru sends her silver, and the gold of California and Australia flows to London; the Chinese grow tea for us, and our coffee, sugar, and spice plantations are in all the Indies. Spain and France are our vineyards, and the Mediterranean our fruit-garden; and our cotton-grounds, which for-merly occupied the Southern United States, are now everywhere in the warmer regions of the earth."[5]

Today, 150 years later, over 60 cities with populations larger than 19th-century London exist around the world. They compete with each other, and all the rural areas, for the use of the planet's natural cap-ital. A city that can offer a comparable quality of life with a smaller Footprint per resident is less dependent on imports and therefore more competitive in the long run.

Astonishingly, current urban Footprints vary considerably. The Footprint of the resident of a compact, medieval Italian city may only be about ⅓ of the resident of a sprawling North American city built during the car age. North American urban settlement patterns are characterized by expansive suburbs, many of which can only be reached by car. Historical cities, for instance in Europe or in the Mediterranean countries, were not designed with cars in mind. They have greater density, are more pedestrian-friendly, functionally more integrated, and typically have better public transit (with trams and buses) than their North American counterparts. The Italians' fondness for fresh, seasonal, and locally grown food not only benefits their cuisine and their health but also contributes decisively to a lower ecological demand—even without high-tech.

North Americans view Italians in their small apartments and houses as penned in. But for Italians the whole city counts as their living room. Therefore, they live in a spacious home. In turn, Italians view the North Americans as isolated in their suburban houses and their lives as constrained by their property lines. Few North Americans can easily walk to a coffee shop or a bar from their home, or go for a quick stroll on their piazza. In other words, the amount of nature we demand says little about how much we enjoy the lives our consumption makes possible, and not even about how spacious our lives feel. North America is starting to recognize this: walkability scores are becoming an ever more important factor in real estate values.

Settlement patterns in cities shape not only their residents' quality of life but also their economic stability. Cities are now competing across the globe. They fight for creative and entrepreneurial talent as well as for locational advantage. One significant advantage is resource efficiency. The cost of resources affects all cities because all of them struggle to access resources that are traded globally. Large Footprints will increasingly become economic risks.

Eventually—as overuse becomes less possible—we are bound by the one-planet limits. Cities with Footprints that exceed what will be available per person by 2050 are particularly exposed; this is simply because their way of living is not replicable across the world, and therefore threatened. It's just a mathematical truth. If cities in the future want to provide thriving lives for their citizens, it serves them to consider now, with every infrastructure project they fund, how they are shaving off resource dependence while increasing quality of life.

Certainly, when adding more infrastructure that requires (fossil) resources to operate, city administrations are destroying the prospects of their own city's future.

If cities choose to address their resource risks, Ecological Footprint assessments allow them to first identify their resource dependence, and break it down to its components (such as mobility, construction, heating, cooling, feeding, waste management, or water provision). This analysis becomes the baseline for meaningful planning, including setting concrete goals for reducing their community's dependence and vulnerability. Whether we look at natural or financial capital, at the metabolism of the material world or at a balancing of expenditures and revenues, both sides require responsible budget management.

This is the point where the communicative strength of the Footprint comes to play. As multilayered as challenges can be in the real world—from home construction to the planning of an industrial area or of a football stadium—the result is always a single number, namely that of the area necessary for *nature* to provide for this project. The method's transparency allows for successful dialogues about the planning process with partners from industry, administration, politics, and, last but not least, the city's residents.

How City Footprints Are Calculated

The Roadmap
A country's consumption Footprint is calculated by assessing the area needed to maintain their resource demand, accommodate the urban spaces, and absorb their associated waste. Final consumption is estimated by adding imports to domestic harvest from ecosystems, and subtracting exports. Detailed information on those material flows is typically only available at the national level—sub-national entities, such as cities or regions do not create and maintain a complete data set on all those resource flows.

Hence, calculating the local Footprint takes an indirect path, involving four steps:

1. Start with the results for the nation using the National Footprint and Biocapacity Accounts.
2. Analyze, for the national average, which activities and consumption items take up which portion of the overall Footprint. The result of this analysis is called the *Consumption–Land-Use Matrix.*
3. Identify the difference in consumption patterns in the city compared to national averages.
4. Use this information to adjust the national average Footprint to local specifics.

These four steps give you a baseline Consumption–Land-Use Matrix for the city for the given year the data were collected.

To make the assessment time sensitive, and also reflective of recent changes, Global Footprint Network recommends using local data to trace changes over time, starting from the baseline year's matrix. We will explain below.

The Consumption–Land-Use Matrix

To assign area demands to human activities, a *Consumption–Land-Use Matrix*—my colleagues call it affectionately the "CLUM"— needs to be established. Because of the complete national data set, we start at the national level. The matrix can be established through manual allocations based on supplementary consumption statistics or using input-output methods. To make these matrices more consistent, Global Footprint Network uses a multi-regional input-output set (GTAP) to calculate these matrices for various countries. These input-output tables provide data on the financial flows between, in GTAP's case, fifty-seven sectors. By overlaying these financial flows with resource intensity, and using some computer power to solve the linear algebra problem, it allows us to attribute resource demands to their respective, specific human consumption activity. In other words, we can calculate how much carbon, cropland, pasture, forest, and built-up land it takes for each particular activity. The main consumption categories are Food, Housing, Mobility, Goods, and Services—and subcategories; this analysis can distinguish about thirty-seven consumption categories. In addition, each one of these consumption categories then can be split into three portions:

1. Short-lived consumption directly paid by households (such as bread, paper, or socks)
2. Short-lived consumption paid for by governments (such as school milk, or government services such as policing or national defense)
3. Long-lasting consumption such as houses, roads, factories— whether paid for by households, government, or firms. The fancy name of this category is *gross fixed capital*. (This term may come handy if you need to impress somebody at a cocktail party.)

City Estimate

The city's Footprint can then be determined by comparing consumption patterns between the city and its host country. Key ratios are based on statistics that are available on a local and national level—such as household size, household income, household expenditures and local purchasing power, distance travelled per person on the road or by train, electricity consumption, food habits, waste generation. Any city's Footprint can be calculated in this way. The more comparative data is available, the more detailed the results.

Tracking Footprint Changes over Time

With this initial *Consumption–Land-Use Matrix* for the city, for one given year, we have a benchmark. We can use it as a reference for calculating changes over time, using just local data. The more local data we have, the more detailed the city trajectory can be calculated. We can identify how local data points affect any cell of the matrix, and adjust the larger pattern accordingly. Through this technique, we can bring the matrix all the way into the present, even identify how particular projects affect the overall performance of the city. With local data, the assessment becomes a valuable tool for mirroring the changes up to the very present.

Uses for This Analysis

Detailed and timely Footprint assessments help cities identify gaps and strengths in resource demand and compare their situations with global, national, and regional trends. Such assessments also highlight priority areas for urban policies, contextualize planned and implemented projects with respect to their resource implications, and create a common language among politicians, administration, and the public. It ultimately allows citizens and administrators to test to what extent the city and its projects are becoming one-planet compatible.

Application Standards

Standards enable analysts to produce consistent and comparable results across applications, avoiding confusion. For this reason, Global Footprint Network and its partners have developed simple standards on how to apply the Footprint methodology in urban areas.[6] Trusted sustainability metrics depend not only on the scientific integrity of the methodology but also on consistent application of the methodology across analyses. Communicating results of analyses must not distort or misrepresent findings. For this reason, Global Footprint Network's standards also provide guidance on communicating results.

The resource strain on competitiveness has already come to the fore in numerous decisions. For example, since the 1990s, Greece, Spain, Italy, and Portugal have heavily increased their resource consumption. This is the consequence of low-cost loans by the European Union that allowed those countries to expand their roads, airports, and housing stocks, which in return invited more resource demand. While these countries have smaller gross national products per person than northern European countries, it led them to run even larger biocapacity deficits, larger even than other European countries. Their resource deficits, amplified by rapidly increasing resource prices in the early 2000s, translated into an ever bigger economic burden on the countries. This growing cost factor plays out like an additional international tax on the economy, slowing it down.

Some have pointed to the Greeks' uneven or unreliable habit of paying their own taxes as being the cause of their economic quagmire during the financial crisis. However, the erratic tax discipline is not a new phenomenon that suddenly occurred or increased in 2008. What was new and may have tipped the scale was a new resources situation: the additional pressure of sky-rocketing resource costs multiplied by a large ecological deficit. Initially the Greek government muddled through with growing financial and resource deficits until that was no longer an option. The rest we read about in the papers. Or you can see the downfall in the Footprint and biocapacity country graphs for all those countries that experienced a similar economic stress: Italy, Spain, Portugal, Ireland all showing similar resource contractions after 2008. The repercussions continue to cause turbulence and suffering in all those countries.

To illustrate the city situations, Global Footprint Network compared the Footprint of nineteen cities in the Mediterranean region. The summary results are shown in Figure 2.1 below. Athens sticks out: with about ⅓ of Greece's population, it requires a biocapacity 20% larger than that of the entire country of Greece.

Resource trends are already key factors determining our economies. We will discuss more in our analysis of Northern Africa and

Figure 2.1. The Footprint of select Mediterranean cities compared to the biocapacity of their home countries. Countries are scaled proportional to their total biocapacity in global hectares (gha). Cities based on 2010 data, countries on 2016 data. Credit: Global Footprint Network. "How can Mediterranean societies thrive in an era of decreasing resources?" 2015. footprintnetwork.org/med.

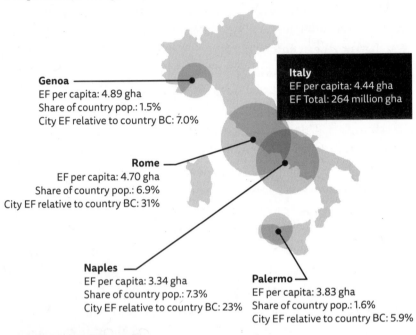

Italy
EF per capita: 4.44 gha
EF Total: 264 million gha

Genoa
EF per capita: 4.89 gha
Share of country pop.: 1.5%
City EF relative to country BC: 7.0%

Rome
EF per capita: 4.70 gha
Share of country pop.: 6.9%
City EF relative to country BC: 31%

Naples
EF per capita: 3.34 gha
Share of country pop.: 7.3%
City EF relative to country BC: 23%

Palermo
EF per capita: 3.83 gha
Share of country pop.: 1.6%
City EF relative to country BC: 5.9%

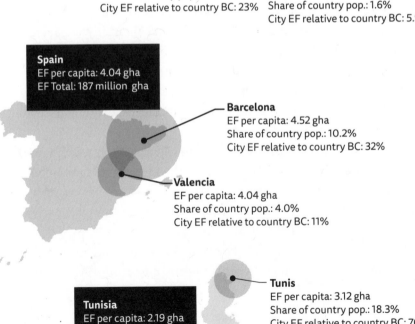

Spain
EF per capita: 4.04 gha
EF Total: 187 million gha

Barcelona
EF per capita: 4.52 gha
Share of country pop.: 10.2%
City EF relative to country BC: 32%

Valencia
EF per capita: 4.04 gha
Share of country pop.: 4.0%
City EF relative to country BC: 11%

Tunisia
EF per capita: 2.19 gha
EF Total: 25 million gha

Tunis
EF per capita: 3.12 gha
Share of country pop.: 18.3%
City EF relative to country BC: 76%

City Footprints: An estimated 80 percent of the world's population will live in urban areas by 2050. In many Mediterranean countries, one or two major urban centers already are major contributors to the national Ecological Footprint (EF) and also run significantly higher per capita Footprints than the average for their nations. Cities thus offer another major opportunity for the Mediterranean region to manage its resources more sustainably, by focusing on drivers and leverage points. The size of each circle on this page reflects the total Ecological Footprint of each city. The size of each country represents its total biocapacity (BC). Although Egypt extends over more area than Turkey, for instance, its size on the map is smaller because Egypt, with its large desert areas, has only about one-third of Turkey's biocapacity.

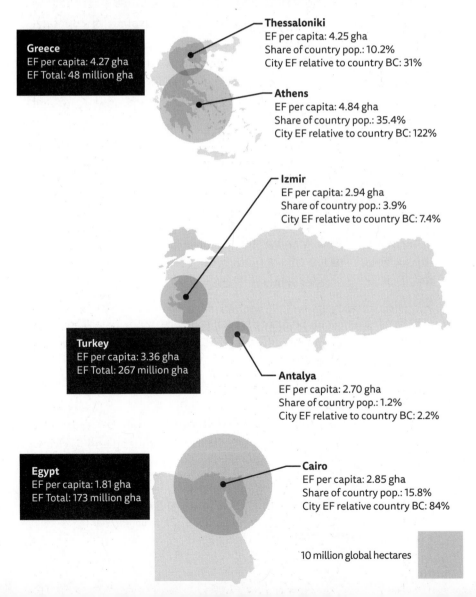

Greece
EF per capita: 4.27 gha
EF Total: 48 million gha

Thessaloniki
EF per capita: 4.25 gha
Share of country pop.: 10.2%
City EF relative to country BC: 31%

Athens
EF per capita: 4.84 gha
Share of country pop.: 35.4%
City EF relative to country BC: 122%

Izmir
EF per capita: 2.94 gha
Share of country pop.: 3.9%
City EF relative to country BC: 7.4%

Turkey
EF per capita: 3.36 gha
EF Total: 267 million gha

Antalya
EF per capita: 2.70 gha
Share of country pop.: 1.2%
City EF relative to country BC: 2.2%

Egypt
EF per capita: 1.81 gha
EF Total: 173 million gha

Cairo
EF per capita: 2.85 gha
Share of country pop.: 15.8%
City EF relative country BC: 84%

10 million global hectares

the Arab Spring in Chapter 4. Yet, despite the realities of increasing resource constraints, most societies do not respond. They behave as if the resulting economic situations are merely cyclical problems and not symptoms of tightening ecological constraints.

Consideration of resources will support better investment decisions. For instance, they help us evaluate which options are more beneficial infrastructure investments necessary for buildings, roads, railways, bridges, or ports. These structures have a lifespan of many decades. A highway system, for example, not only increases our current dependency on fossil energies but also solidifies it far into the future. It pours, so to speak, our Footprint in asphalt or concrete. Changing directions later will become costly and tenacious. This is how investment in the wrong kind of infrastructures becomes a trap; some call those ill-conceived investments *stranded assets*.

Let's take an imaginary city exploration. Travel downtown and find yourself a comfortable place to linger.[7] Look around. The people, buildings, and cars all look familiar. Now adjust your perspective and trace the presence of fossil fuel. Of course, any lighting, heating or cooling, and transportation relies on it, as do water pumps, elevators, and an immense number of household and office gadgets. In addition, many items require energy without it being instantly obvious as you look around. Making concrete takes a huge amount of fossil fuel-powered energy. The same is true for the glass panes of shop windows or the steel of vehicles. The omnipresent plastics, from litter floating through the street to furniture, water bottles, or carpets, is another reminder of how much fossil fuel surrounds us. We are taking it for granted, and do not even see it any longer. But the Footprint does not overlook it.

And for good reason. As already mentioned, 60% of humanity's Footprint is caused by our use of fossil energy. And the largest part of that fossil energy is consumed by urban centers, their buildings and their transportation systems, most of which begin and end in cities.

Food products in North America, for example, typically travel more than 1,500 miles from the farm to the fork.[8] Industrialized agriculture is energy-intensive in every way imaginable. Tractors run on diesel. Artificial fertilizers are made from fossil gas. Pesticides and herbicides are synthesized from oil. Agricultural products are shrink-wrapped in plastic, are cooled and then heated, mostly with fossil energy. Every calorie of food offered in the cities' supermarkets uses up on average ten calories in fossil energy for its production, distribution, and preparation.

Large quantities of energy are spent in the morning when people travel to work and in the evening when they return home.[9] Suburbia, the dream of living in the country and working in the city, has devoured vast expanses of cropland and forest. Over the past decades, many cities have mushroomed. Rapid urbanization is staggering particularly in low-income regions. Megacities like Mexico, Laos, or Bangkok continue to expand with densely populated, poorly served informal settlements. At the other end of the spectrum, high-income cities experience urban sprawl with ever larger homes. In such cities, spatial expansion markedly surpasses population growth. In the US, for instance, from 1950 to 2010, urban land more than tripled while the urban population doubled.[10]

Los Angeles, an immense agglomeration with 17,800,000 residents, is famous for its gigantic highway system and overpasses. The vast majority of residents take their car to work. Greater London, on the other hand, has a significantly higher population density with no fewer than 14,700,000 residents.[11] The British capital with its typical semi-detached and row houses in its suburbs is in turn several times more spread out than Hong Kong, one of the most densely populated cities in the world. No wonder Hong Kong manages area (fuel and other resources) much more efficiently than Los Angeles or London. Still, that doesn't mean that denser is always more efficient. Extremely high skyscrapers devour considerable amounts of energy through the elaborate infrastructure: elevators, lighting, water pumping, and more heating and cooling due to their exposure. Six floors, as in much of Paris, is close to optimal. Imagine a Paris with electric rickshaws and scooters instead of cars, bicycles, and walkways, and with energy-neutral architecture that is also designed to integrate residential spaces, work, and leisure.

In 1900, less than 30% of humanity lived in cities, while now more than half do. Since then the global population of 1.5 billion has grown to well over 7.7 billion, with a trend toward 9–10 billion by the mid-21st century.[12] Almost all of these additional people will live in cities, many of them in the slums of the megacities that have more than ten million residents. They will dwell in structures such as the one emerging along a 300-mile axis connecting the two biggest Brazilian metropolises, Rio de Janeiro and São Paulo: an agglomeration of currently thirty-five million people, rivaling the Tokyo-Yokohama region and Guangzhou, both with over forty million people.[13] Cities, the most endurable creation of higher civilization, are in the process of creating ever vaster, more massive structures with more complexity, and potentially more fragility, than we have ever built. The recent water crises in Cape Town or São Paulo emphasize the physical dependencies of these megastructures.

When Manhattan still had a population of 30,000 in the late 1700s, they realized that the island, if built out, could house a much larger population. To feed that larger population, the Eire Canal was conceived and built. When the Canal was opened in 1821, Manhattan's population was just above 100,000 people. Thereafter, it added another 100,000 or more every ten years, reaching 2.3 million people in 1920, just 100 years later.[14] What distinguishes this story is the planners' concern for resource security from the early planning phases of a built-out Manhattan. In contrast, today megacities are built and expanded without reasonable plans for those cities' resource security.

In times of resource constraints, local governments, their administrations, business leaders, and indeed every citizen committed to any city's long-term success will have to ask: how much biocapacity does our city depend on? How can we track where our resources come from, and whether these places will be able to? What are they used for? Where does our waste end up?

In essence, postponing forward-thinking action and continuing to promote resource-inefficient infrastructure will cause lasting liabilities. The core question becomes: how can we reduce our city's metabolism? And more questions unfold if we want to succeed in better positioning our city so it can handle the global competition. How can we protect our local natural capital—water, for example—to use it sustainably? How can we make progress toward a resource-light city and reduce our dependence on imported resources? How can we create an infrastructure for mobility, water, and electricity that will not turn into an ecological trap one day but is efficient and enables comfortable living?

Technology can be our friend. With today's technologies, cities with a relatively big Footprint can reduce their resource requirements by a factor of five,[15] for example, by curbing the energy required by buildings, sourcing renewable electricity, and reducing mobility needs by bringing work, living, and shopping closer together. The Footprint provides metrics for all of the above questions.

CHAPTER 3

CROPLANDS, FORESTS, AND OCEANS

How Much Biocapacity Do We Have?

Every forest is a stock of trees. If we take some out, we initiate a re-source flow in the human economy: trees become timber or fuel. During the Industrial Revolution when people started to burn coal, oil, and gas on a large scale instead of wood, these supplies were ini-tially immense. First, there were the coal seams (the "subterranean forests"), later the oil and gas fields. And so the energy flows contin-ued to grow. With improving access to the supply, more uses for the energy fueled the expansion. With decreasing effort, including less energy and fewer human hours, we have been able to extract great quantities of energy. Harvesting more energy by expending less en-ergy is of course seducing us into ever more consumption. But by now it has become obvious that the combustion residues, not least carbon dioxide, are a problem that are limiting the utility of these energy sources, even more limiting than the fossil deposits underground. It takes forests and oceans to absorb greenhouse gases, at least as long as humans don't produce machines that capture and store all the car-bon dioxide. So far, we are not doing so. One reason is that current carbon sequestration technologies are still laborious, expensive, and technically immature. Also, it always and unavoidably will take a lot of energy to capture that CO_2. As a result, human beings, despite all their technology and innovation, remain dependent on the bio-sphere, the planet's living surface. In fact, with or in spite of all the technology, the dependence has increased. Technology has allowed us to do more, but usually by also using more resources. Footprint accounting shows the size of nature we now claim, compared to what we actually have.

Every single astronaut who has travelled in orbit tells us of the great awe they experienced when they looked down on Earth. They were impressed mainly with the planet's beauty, but also with its vulnerability. They often talk about the layer of royal blue horizon that spans the Earth. That is our atmosphere, the layer of air that protects us from dangerous radiation, carries water into the mountains, balances the surface temperature, and supplies us with oxygen. Seen from a distance, the atmosphere is unbelievably thin and has an even more delicate barrier. Beyond Earth's atmosphere, the deep, magic black of space opens up. If we watch the planet's surface over time, we can see how the seasonal waves of vegetation on the ground, driven by photosynthesis, roll like ebb and flow across forests, steppes, cropland, and grazing land. Earth is a miraculous, self-regulating system that has developed over four billion years.

On July 23, 1971, the first functioning ERTS (Earth Resources Technology Satellite), called Landsat, set out into space. Its orbit had an altitude of 570 miles. Every 16 days, it flew over the same location on Earth. Landsat was equipped with high-quality color cameras that make it possible to identify different vegetation zones and ecosystems.[1] Chlorophyll, for example, reflects less than 20% of longwave, still visible light and about 60% of radiation from the infrared spectrum. Since the 1970s, satellites have continuously improved. Currently Landsat 7 and 8 are doing their rounds—Landsat 10 may join in late 2020. Most of the images for online map portals, such as Google Maps, also come from this civilian Earth observation satellite run by NASA. By now every location on the planet has been measured and mapped, with the photographs' resolution today amounting to a few meters. Thanks to the technology of satellites, cameras, and computers, we have gained a deeper understanding of the planet's surface, of its details, as well as of its ecosystems' interactions and hence of the biosphere as a whole.

Satellites contribute data for Ecological Footprint accounting. The information they collect first goes to the countries' statistical offices, which analyze areas and land use. They pass on the data to the statistical bureaus of the United Nations. These United Nations data, which allow for global comparisons, are then used by Global Footprint Network for its calculations.

Based on these data sets, National Footprint and Biocapacity Accounting distinguishes between different types of landscapes. In the physical reality, there can of course be blurriness and transitions. For example, the official statistics the Footprint draws on include very sparse tree populations still under the category of "Forest" even though those are almost steppe. This is one of the reasons the Footprint is overall rather conservative and deliberately describes a situation more optimistically than it really is.

The same is true for the yields of industrialized agriculture. The existing numbers, which come from the United Nations' Food and Agriculture Organization (FAO), do not tell us how much of the higher yields results from the massive use of fossil energy and agrochemicals. Or where a soil is degraded long-term as a result of erosion. Or where groundwater reserves are overexploited. In the long run, maintaining current yield levels will hardly be possible.

Footprint accounting for nations counts as biocapacity whatever is produced each year. If current methods of production degrade ecosystems, the accounts will show evidence only when the degradation has already occurred. The Footprint does not offer forecasts; neither does financial accounting. Instead of speculating about the future, the Footprint documents historical trends and gives managers a stable baseline for decisions.

Thanks to the official data sets of the statistical bureaus of the United Nations, we have a consistent base for our Footprint calculations. Each year, new and current data are added, and historical data sometimes even corrected. That way, the ecological accounts of each country can be compared over time with those of each other country or the planet as a whole.

The Footprint is a metric that places human consumption of natural resources and services at the center. It asks how much biocapacity is used in our economic processes. It builds on ecologically informed methods that recognizes regeneration and consumption. It draws on assessments of *net primary productivity*,[2] with whose help researchers describe the annual biomass regeneration of ecosystems.[3] But Footprint accounting makes the measure sharper and simpler by taking a more agricultural measurement perspective. It does this by focusing on mutually exclusive areas needed to provide for human demand. This avoids a more speculative approach, one that compares how much biomass is removed to how much is being regenerated. Ecological Footprint accounting merely add the bits of area that are used exclusively to provide for particular demands such as fruits, vegetables, grazing, or timber.[4]

Another simplification is that the Ecological Footprint analysis focuses only on those areas that are biologically productive. Marginal areas that overall contribute insignificantly little to the human metabolism are left out of the analysis. For instance, the open seas contribute only a small portion of the global fish catch. Therefore, they are not included in Footprint calculations, and neither are deserts or regions covered in ice. On the other hand, coastal waters, continental shelves, areas with nutrient-rich deep-sea currents, and marshlands or river deltas—which together support 90% of fishing yields—are included.

Highly productive ecosystems, for example, in temperate zones, regenerate relatively quickly. In alpine areas or in the tundra, on the other hand, vegetative processes are slower and more susceptible to disturbance. Dry grasslands, as in Australia, cannot be used for intensive grazing. Cattle cannot reach all the available biomass for lack of water holes. If cattle move too far away from a water hole, they die of thirst. Also, these areas can be fragile, with the weight of cattle destroying soils, and turns such pasture operations into "mining."

The smaller the biocapacity of an area, the less it can be used. It takes more effort to concentrate its output, and often these areas are also more fragile, tending to easy degradation. If productivity drops below a certain threshold, the usefulness of the area dramatically diminishes, and in most cases, such an area stops yielding harvests of any use to humans at all. The energy return to energy investment is just too unfavorable. Therefore, Ecological Footprint accounting limits itself to the areas that are biologically productive, and excludes the marginally or non-productive spaces.

Currently the sum total of all biologically productive areas of water or land on Earth amounts to about 12.2 billion hectares or 47 million square miles. This is close to ¼ of our planet's entire surface.[5]

With Footprint and biocapacity accounting, researchers at Global Footprint Network have developed a method to capture the multitude of ecosystems. The national accounting method distinguishes between five different land types:

1. Cropland has the greatest bioproductivity per hectare. In Footprint calculations, this land type represents the sum of all harvestable field crops, such as corn, oily fruits, cotton, and many others.

2. Grazing land supports animals that produce meat, milk, or wool. Animal products, however, do not rely exclusively on grazing. For example, cattle are often fed crops such as soy or corn.

3. Fishing grounds are counted proportionally to the maximum sustainable fishing yields each area allows. This applies to both lakes and coastal waters. Increasingly, fish is also produced in ocean-based farms, fed with lower-grade fish and products from other land types (e.g., soy from cropland).

4. Built-up land was typically quite productive before converted to urban uses. Most cities are built where agriculture was easy; exceptions are new cities such as Las Vegas or Dubai, or port cities like San Francisco or Goteborg. Urban areas retain some biocapacity thanks to gardens, roadside greenery, and the like. Home and road construction has reduced the biocapacity that otherwise would be available for production. Built-up land hence reflects agricultural potential that was relinquished in favor of cities, towns, villages, and roads.

5. Forests provide for two competing Footprint demands. One delivers all the wood and fiber which forests produce as lumber or

for paper production. But if managed appropriately, other forests can also provide for the second demand: they can absorb carbon dioxide from fossil energy. Harvesting forest products and absorbing carbon dioxide from fossil energy are (largely) two mutually exclusive functions.[6] When wood is harvested, the stored carbon dioxide will sooner or later be released again. To store carbon dioxide for good, trees have to remain in place. And if the absorption of carbon dioxide is our goal, we need to know how much forest is required long-term. How much forest have we already set aside and legally secured for this purpose? This portion of the forest land-type services the Footprint of the CO_2 emissions of fossil energy. But because we do not know how much forest has actually been protected legally and long-term for CO_2 sequestration, National Footprint and Biocapacity Accounts record forest as one single category in the biocapacity accounts. They do not split it up for its respective functions. On the Footprint side, the accounts, however, distinguish between the forest-products Footprint and the carbon Footprint.

Let's imagine a farm with overfertilized fields. Rain will wash excess nutrients into a creek. The creek flows into a lake beyond the farm. Slowly, the lake gets turbid. As it takes on ecological service functions for the farm, the lake becomes a sink for waste and produces fewer fish. In this way, the farm is similar to a state that emits carbon dioxide, expecting some ecosystem somewhere else in the world (i.e., the lake from the above analogy) to cope with it. Of course, the people from the lake would and should complain to the farm. They would visit the farmer of the overfertilized field and push her to change her practices and pay for the damage to the lake.

For most CO_2 emissions from fossil fuel, no land is set aside to absorb it. Today, only a minute portion of CO_2 from fossil fuel falls under trading schemes or has any price attached. Even the scientific consensus body, IPCC, recognizes and warns that our current trend commits us to at least 3°C warming. IPCC points out that carbon emissions would have to be cut to about half by 2030 from current

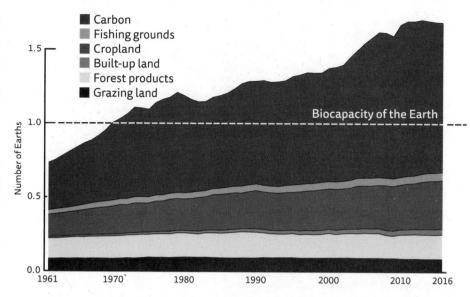

Figure 3.1. Humanity's Footprint, historical trends, by different area uses in number of Earths required. Credit: Global Footprint Network—National Footprint and Biocapacity Accounts 2019 edition, data.footprintnetwork.org.

levels to give the Paris climate goal a fair chance.[7] And cut at least in half again every decade, and soon move to net negative emissions.

This data makes it doubly absurd that it is still acceptable today to emit CO_2 with no restraint and no duty to pay for its absorption or to set adequate forest areas aside. Instead we seem to implicitly hope that the service will be provided "elsewhere" on our behalf. There is no infinite sink for our waste flows; hence our carbon waste accumulates in the atmosphere and the oceans; this is a rock-solid ecological reality.

The core purpose of the Ecological Footprint is to create a consistent, aggregated indicator. For that it needs a common unit of measure that can consistently compare biocapacity and demand, even though productivities of ecosystems vary widely. This is resolved by using average hectares, with *average productivity* as the measurement unit.

The National Footprint and Biocapacity Accounts resolve this translation from actual hectares to average hectares using, as a first

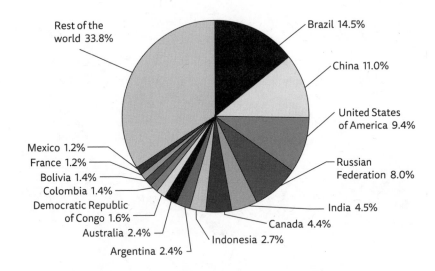

Figure 3.2. Distribution of global biocapacity by country (in percent of global biocapacity). Credit: Global Footprint Network—National Footprint and Biocapacity Accounts 2019 edition, data.footprintnetwork.org.

approximation, yield and equivalence factors within each area type and each year.[8] *Yield factors* compare productivities of similar areas between countries. For instance, they describe how productive cropland is in France compared to cropland in Colombia or the world as a whole. *Equivalence factors* convert types of land into average land with world-average productivity. In other words, they compare, for instance, how much biocapacity one hectare of average forest has compared to one hectare of average cropland. By multiplying a given hectare with its yield and equivalence factor, we get an estimate of how many world average hectares this particular hectare embodies. We call those average hectares *global hectares*.

Global hectares are the Footprint's central unit of measurement. The global hectare is an area of 100 meters by 100 meters (10,000 square meters) of biologically productive Earth surface with average global productivity. A hectare is about 2.47 acres or 108,000 square feet. We have, as previously mentioned, roughly 12.2 billion global hectares of biologically productive surface areas on Earth represent about ¼ of the planet's total surface area.

On these biologically productive areas, photosynthetic activity takes place. With the continual supply of solar energy, plants use carbon dioxide to release oxygen and build biomass. Photosynthesis is thus the basis for all animals' food chains. It is the engine powering the animal kingdom. The associated food chains represent the energy flows (or the metabolism) of all animal life in the biosphere.[9] No other process has shaped nature's evolution as much as photosynthesis. It is photosynthesis that crucially contributes to the development of the atmosphere. Thanks to photosynthesis, which took possibly about one billion years to emerge after the creation of Earth over 4.7 billion years ago, the planet's surface has transformed itself throughout its existence from an inhospitable place into a self-regenerating and self-regulating system with a great diversity of living beings.[10] Without this magical bio-machine, planet Earth would be as lively as Mars.

Admittedly, over the course of its history and particularly the last 1,000 years, humanity has thoroughly "tidied up" nature. By now about ½ of once untouched forests have been converted into grazing land, cropland, and urban areas. On all continents, human beings have driven many of the large wild mammals from their biotopes, even eradicated them (Africa is something of an exception).[11] This is particularly true for land-based predators such as tigers, bears, and wolves.

The higher an animal in the food chain, the rarer it is at this time in human history. It takes plenty of biomass to feed an animal one step up in the food chain.[12] Each step up the ladder takes a creature further away from the original production of photosynthesis, and exponentially more energy is needed to maintain this animal. Also, any variation in the productivity of ecosystems has a major impact on animals higher up the food chain. In the tundra, the level of photosynthesis is low due to cold temperatures and lack of sunshine. Besides, in those harsh climates, animals have to build food depots to hibernate. Carrying capacity for animals is defined by the food availability during the toughest time of the year, further reducing animal density. As a result, fewer creatures can survive in that area. The biodiversity is smaller and the food chain simpler. At the opposite end are forests in the temperate and, even more so, in the tropical climate zones. They produce distinctly more biomass. Their ecosystems are richer and more diversified.

By no means insignificant are the parts of the Earth's surface that have been declared conservation areas, a total of 20 million square kilometers of terrestrial areas (i.e., 8 million square miles or 14.7% of terrestrial area) and 15 million square kilometers ocean areas (6 million square miles or 10% of the national coastal areas), although a large portion of this land is still *de facto* impacted by people.[13] Year after year national parks such as Yosemite Park in California are populated by millions of visitors with camper vans and tents.

Man has subdued the Earth, as the Christian Bible puts it. From being the rare species that in its early days was itself often threatened with extinction, *Homo sapiens* has claimed absolute dominion over the rest of nature and has essentially domesticated the planet's entire biosphere. Most rivers have been regulated and channeled, often with catastrophic results. So much water is taken from the Nile in Egypt that for large parts of the year it no longer reaches the sea. Similarly, the water of the Rio Grande gets overused so much that by the time it reaches Mexico it is so high in salt it is no longer good enough for agricultural use. Eventually the river simply seeps away somewhere in the landscape, only to reemerge several hundred miles later via a tributary river into a new, modest life.

China's Yellow River has already been diverted and diked many times during its history, but today its very existence appears threatened. Its water level and flow rate have been reduced so much that the silt it carries is now being deposited in the river system itself and year after year the Yellow River breaks out of its levees. The near total draining of Lake Aral was the cost and consequence of the work of Soviet hydraulic engineers to irrigate water-thirsty cotton fields.[14]

The romantic notion that everything in its original state is somehow good and everything artificially created is somehow bad no longer makes sense. The last untouched ecosystems are so completely surrounded by cultivated landscapes that true wilderness no longer really exists.[15]

Today, human beings move more mass (soil, biomass, minerals) than the natural forces of wind and water together. This metabolism (the material exchange between human beings and nature) has grown in more than quantity. History is a succession of qualitatively different metabolic systems. As hunters and gatherers, humans consumed about one ton of nature per person per year for food, basic housing, and weapons. In agricultural societies, the level rose to three to five tons. It was limited by a lack of energy. In industrialized societies, finally, the consumption of nature per person per year is around fifty tons, plus our consumption of water and air.[16] And populations are now far higher too.

In agricultural societies, farming forms the dominant resource base for the economy. It produces not only food but also fiber (wool, hemp, and flax), as well as oils and dyes, and finally furs, leather, and bone. Through biomass, mostly wood, farming was also the main source of storable energy. That, in turn, was the prerequisite for the production of mineral raw materials, such as salt, ceramic, metals, or bricks.[17]

Producing wood required far less labor than growing grain. People therefore often thought of wood as a free product they could harvest but didn't need to plant. But when wood became scarce—a common phenomenon in preindustrial times, before the discovery of massive fossil energy reserves—it became clear: logging "destroys" the forest the same way harvesting "destroys" the wheat field. Producing biomass requires area. And any area is finite. Different kinds of land use typically compete with each other: it is either/or.

To build an average ship in the 18th century, the timber of 20 hectares of high forest was required (about 2,000 trees each of which weighed about two tons). It then took 50 to 80 years to regrow that stock of trees. One ship hence required 20 hectares for 50 to 80 years. No wonder that, at the time in England, the mother country of industrialization, large swaths of forest were logged for shipbuilding alone. Older and larger trees became rare. While material for planks continued to be produced inside the country, the timber for masts increasingly had to be imported from Scandinavia and Russia.

Another example of the relation of forests to industry is the "Saline Royale," the French salt factory set up under Louis XV in Arc-et-Senans close to Dole. It was supported by the Forêt de Chaux, a forest of 80 square miles (20,000 hectares) to provide the fuel for boiling brine and turning it into dry salt. That forest, in essence, was the Footprint for the salt factory.

Agrarian societies were completely dependent on solar energy. Only later, in agro-industrial systems, fossil fuel dependent inputs, such as artificial fertilizers or diesel for tractors, were added. Until that happened, dependence on area size and sunlight was absolute. When wood became scarce, it had to be imported, or people had to re-forest, in which case that land could no longer be used to grow grain. To use flowing water, often dams had to be built; these dams flooded land that previously had been forests, or for grazing, crops, or settlements.

Fishing significantly improves the menu. Those living close to water could supplement their access to protein from fishing rather than from animal husbandry or hunting. For fishing nations such as England or the Netherlands, inland areas not needed to produce protein became available for other usage. In this sense, the vastness of the seas is clearly part of the resource base in coastal areas.

The basic rule is this: people in agricultural societies based entirely on solar power relied on area to manage the flows of materials and energies. In such a system, any natural stocks were small. Even forests that stored biomass over several decades could not be consumed beyond a certain point or would be destroyed. In an agrarian solar-energy system, energy density and energy flows remained limited. And when energy is scarce, everything is scarce. In such societies, farm management and running countries is quite similar: managing their agricultural potential poorly translates into poverty and hunger.

All of that changed rapidly when people discovered the large-scale use of coal, oil, and gas. With the beginning of the industrial age, people figured out how to open the vaults of solidified solar energy from the past and suddenly could draw on seemingly infinite energy depots. From these they were able to generate an enormous flow of energy. This energy flow stimulated ever more new uses for this energy, growing its demand exponentially.

In short, with the transition from solar-based agricultural to industrial societies, humanity turned a crucial corner in its development. *Per capita* energy consumption significantly increased, and population density kept growing. For the time being, resource restrictions had been overcome, the constraints through area size shattered.

The fossil energy system especially revolutionized mobility and transport. Through millennia, transport on land had been arduous, slow, and expensive. Only ships enabled the transportation of larger loads. With the arrival of industrial society came railways, and in the 20th century, mass mobilization through cars, trucks, and planes. The significance of horses shrank drastically. For centuries, they had transported people and goods or pulled plows. Today, they are used for riding and breeding but mostly as toys for young girls, and to animate betting games. The share of grazing land used by horses and draft animals has fallen. What has gone up instead are the numbers of roads, highways, and parking lots for the herds of cars moving across the planet.

Energy flows go hand in hand with material flows, as more energy allows economies to transform more materials. As a consequence, the flows of materials and goods have swollen since industrialization.[18] Technological, logistic, and political developments have also contributed. The trend continues in our globalized present.

The first container ship, a converted former tanker, set out to sea in 1956 with 58 standardized metal crates. Today the carrying capacity of the largest freighters is 20,000 units. These gigantic container ships, built from eight times as much steel as the Eiffel Tower, stay in port for a mere few hours. Loading and unloading is almost entirely automated. And off they set again.

This understanding of our natural capital's stocks and flows that we can observe in the transition from agricultural to industrial society also lies at the heart of the Ecological Footprint. Footprint accounting does not just ask how many tomatoes I will eat this year.

It takes it a step further to inquire how much garden or crop area is necessary to produce the number of tomatoes I will eat. The Footprint keeps account of what the Earth produces in a given period of time. Footprint accounting remains anchored in the description of area types and their production. While the method cannot comprehensively describe the stock of an entire ecosystem—such as a forest or an ocean—it tracks consistently income and expenditure, measured in physical global hectares. This sheds light on whether we are building or depleting natural capital, and shows the direction in which it is developing.

A forest, for example, contains a measurable amount of biomass. Nature continuously tops up that amount, as is clearly visible in the growth rings of trees. By assessing biomass amounts over time, foresters know how much wood they can take without weakening a forest. The details of stock and flow are quite complex, but the essence is simple, just like a water reservoir with inflows and outflows. The account balance is visible from the reservoir's water level. Footprint accounting builds on the same stock and flow principles. The Footprint answers how much biocapacity is necessary in order not to decrease capital stock.

A mature forest contains roughly as much biomass as is produced over fifty years. Most ecosystems, on the other hand, maintain a significantly lower stock. The oceans' entire biomass represents eleven days of the entire ocean's production.[19] Things are very different with fossil deposits, which form a gigantic stock built over enormous periods of time with deposit rates possibly being a million times slower than current exploitation.[20] So fossil fuel stocks keep shrinking.

The great dream of fossil fuels promised infinite energy at low cost. After World War II, the industrialized world lived through a truly golden age with enormous economic growth rates.

In Europe, North America, and Japan, broad sections of the population experienced a level of prosperity that during their grandparents' time had been accessible only to the few millionaires. Telephones, refrigerators, permanently heated (or cooled) living rooms or bedrooms, private washing machines, and cars proliferated. From 1949

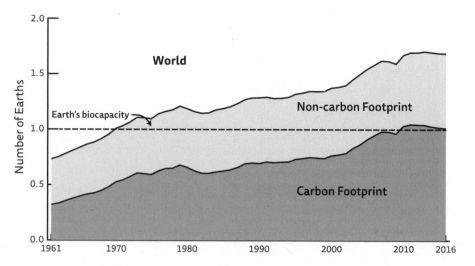

Figure 3.3. Humanity's Footprint by carbon and non-carbon. It shows the proportion of Earth's biocapacity needed to neutralize carbon dioxide from fossil fuel burning. In number of Earths.

to 1972, worldwide energy consumption tripled.[21] At the same time, oil kept getting cheaper. The more we burned, the more deposits were discovered—one huge energy rush. Only the 1973 oil crisis dampened our exhilaration. Still, our consumption of fossil energy did not diminish—on the contrary. This even extended to our use of fibers: while 97% of the fibers used in the early 1960s were biological, they make up barely ⅓ today.[22]

Solar-based agrarian societies were defined by resources' and energy's *limited flow*. Since the beginning of the fossil age, these flows have exploded in size—until we realized that we are also defined by *limited stock*. During the past 200 years, humanity has shoveled unimaginable amounts of carbon from the Earth's crust into the atmosphere, with the result that its carbon dioxide level has increased from 278 parts per million before the industrial revolution to about 411 ppm in 2019.[23] The liberation from the tight limitations of solar-based agrarian society, in particular its dependence on area, has become an irony of history. Again area is the limit: oceans and forests are overburdened by the task of absorbing gigantic amounts of carbon dioxide. This limit is far tighter than the limits on fossil fuel stocks underground.

If the planet's biocapacity were used to absorb all of the carbon dioxide we are producing, we would not have a sufficiently large area left to produce wood or corn or potatoes. The Footprint looks at the whole biosphere, not just at the climate. It describes the biosphere's current state and identifies the "guardrails"[24] of sustainable development. Footprint accounts identify the natural supply of *biocapacity*, measured in productive land and water areas relative to demands put on it by humanity. These accounts give us a statement of our global ecological account. Now we can choose how to act. As with literal bank statements, we can read them and draw conclusions, or toss them unopened into the wastebasket.

In Summary: How Much Nature Do We Have?

No more dilly-dallying. All put together, what is available to accommodate humanity's Footprint? Obviously, nature is everywhere, but life is concentrated where the ground is covered by photosynthesis— the green areas and the oceans close to the coast. That's where biocapacity is located.

How much green space there is on the planet is quite simple to calculate. You may remember from history class that (just after the French Revolution) the meter was defined in relationship to the size of the Earth. To make length "universal," people then declared that the distance from the equator to the North Pole should be 10,000 kilometers; this makes the circumference of our planet 40,000 kilometers (or nearly 25,000 miles).

When they measured out the planet's circumference back then, they did not get it exactly right (they missed seventy-five kilometers when measuring around the equator, or eight kilometers if they measured it via the North and South Pole). But that's close enough. If you dare to take some high school geometry out of the closet, you may remember that the surface of a sphere is $4\pi r^2$. The circumference is $2\pi r$ or 40,000 kilometers. Now, with a little algebraic juggling, you can therefore calculate that Earth's surface measures fifty-one billion hectares.

Looking at vegetation maps, we can find out that roughly ¼ of this surface—forests, cropland, wetland, and fishing grounds—is biologically productive. The rest is deserts, sheets of ice, and open seas that are low in fish. Of course, a more precise way of doing this is to take land use statistics and add up these productive areas. With a global population of more than 7.7 billion people, these additional calculations give us about 1.60 hectares of biologically productive area per person in 2019 (or 1.63 in 2016 when the population was smaller). These 1.60 hectares are in this case also identical to global hectares, because they represent the global average.

Since we compete with millions of species of wild animals for this limited biologically productive space, we may want to leave a good portion of Earth's biocapacity for them. The eminent biologist E. O. Wilson dedicated his most recent book to this question, calling it *Half-Earth*. His vision is to leave half of the planet for wild species. Whether this is generous or not is another question: We are just one species surrounded by a few domesticated species, living in symbiosis with many more such as the flora in our intestines, and some that live off us such as disease bacteria and viruses and other parasites. But the wild species are millions, maybe a hundred million. E. O. Wilson estimates that leaving half the planet for other species would keep about 85% of our biodiversity intact.[25]

Perhaps we want to keep additional reserves because the global population keeps growing? Using all these numbers, you can set and advocate for a specific biocapacity budget goal.

With 3.6 global hectares of biocapacity per person, the US is doubly as rich as the world's average. Canada is even richer with over 15 global hectares per person—that's nearly tenfold the global average. Yet to produce everything an average Canadian consumes and to also absorb the associated waste (especially the CO_2 of fossil energy), one Canadian needs over 8 global hectares—almost five times the size of what is available per person in the world.

The global number of 1.60 global hectares of biologically productive space per person starts as a purely mathematical measure. There is no secret "eco-communist" demand that every individual must be allotted the same share. One point six hectares is not a marker of what is fair, nor does it prove what should be fair. It is just a descriptor of what is, a simple division of how much planetary biocapacity there is by number of people on this planet. It definitely helps orient the fairness discussion. For countries with a *per capita* Footprint higher than 1.60 global hectares, it essentially indicates that their consumption of nature cannot be replicated across the world. Also, this threshold number needs to be set even lower if we recognize the importance of biodiversity.

CHAPTER 4

ONE PLANET

Ecological Limits and Then What?

We catch more fish than fishing grounds replenish; we release more carbon dioxide into the atmosphere than our ecosystems can absorb. In some regions we cut more trees than can regrow, and pump more water from the ground than recharges. This phenomenon is called *overshoot*. Although the phenomenon is dangerous and of paramount importance, the English term has no equivalent in German, French, Spanish, Italian, Arabic, and probably most languages—as if doctors had no name for one of the most important diseases. This linguistic gap shows the double risk of overshoot: not only is the existence of overshoot itself full of risk, but so is the fact that the phenomenon is almost off our cultural radar, in spite of the world economy's intense global ecological overshoot.

People may experience overshoot as an unpleasant side effect of their economic activities. Countries can add to global overshoot without locally overexploiting their resources. If they have the financial means, they can spare their own ecosystems and use imports, or they can simply release their carbon waste into the global atmosphere. This way a country with a biocapacity deficit may not experience overshoot in its own territory.

In addition, overshoot tends to arrive gradually, which makes it that much more dangerous. But from a certain point onward, ecosystems can give in, and lose productivity. We may call them collapsed. They no longer provide what we want from them. The Footprint shows us how much biocapacity we currently have, and how we use it. That way we can see whether we are acting in our own best interest for today as well as tomorrow.

Anyone born in 1950 has witnessed the almost incredible growth of the global population from 2.5 to more than 7.7 billion people today. During the second half of the 20th century, they have experienced a sevenfold increase of the world economy. In the same period, global water consumption tripled, carbon dioxide emissions quadrupled, and fish harvests quintupled.[1] Never before in the history of humanity has there been a comparable growth spurt. Can we keep up this level over the long term?

According to the Footprint, humanity in the middle of the last century used annually about half the planet's biocapacity. Our children, born around the turn of the millennium, had an altogether different start in life.[2] In 2000, the planet's biocapacity was already overexploited. In 2019, we used 1.75 Earths (meaning that we consumed Earth's products 75% faster than they regenerated). If we follow the United Nations' conservative projections about population development and the growing demand for resources, when our children (born at the turn of the century) hit their fifties, humanity would be consuming three times the planet's ecological capacity. This sounds absurd, and it is probably physically impossible.

At the moment of writing, we do have a few reduction targets for carbon dioxide. The global emissions are still on the rise in spite of some local and national efforts such as decarbonization programs in Scotland and Costa Rica. The question still remains open how to decarbonize without putting additional pressure elsewhere, for instance through the increased use of biomass. This makes all the graphs for the future still point upward, including for global population size.

If people 2,000 years ago would have been asked what the world would look like in 50 years, they would have found it easy to answer:

Like today. The speed of change—whether in population development, technological innovation, or communication—was so minimal they would have hardly noticed. Rulers came and went. Nature's challenges, from droughts to floods, were borne with patience.

But for the coming 50 years, a number of existential questions arise. What will be the consequences if a global population of roughly seven and a half billion people develops toward nine to ten billion? What will happen if humanity does not successfully master this transition and bottlenecks occur along the way? Currently industrial lifestyles are growing even more rapidly than world population.[3]

Populous countries with newly expanded economies such as urban China, India, Brazil, or Indonesia are in a tremendous race to embrace and extend industrial lifestyles. From 2000 to 2010, China almost doubled its energy consumption, and in recent years the increase has barely slowed down. By 2050, an additional three billion people are predicted to have joined the high-consumption global middle class—an understandable if almost unimaginable trend given physical reality. Where would the required resources come from? The age of fossil energy will somehow come to an end, regardless of new discoveries of natural gas and fracking technologies because the cost of extracting energy units keeps going up. Extraction will therefore shrink well before "the last drop of oil," whether we like it or not. Partly, it will happen because the reserves will run out or will simply require too much energy to extract. But that limitation may be far too late for the climate.

A more hopeful scenario reduces or abandons extraction to avoid climate damage. The 2°C target that G8 heads of state first acknowledged in 2009 in Copenhagen requires fast and radical reductions of emissions. In June 2015, the G7 heads of state pledged to end the burning of fossil fuels by the end of the century. By December 2015, the Paris Climate Agreement signed by 190 countries asks to limit global warming to well below 2°C, and if possible 1.5°.

Even the 2° target is not achievable without phasing out fossil energy well before 2050. The argument is simple. Consensus-based

IPCC reports from 2014 tell us that 450 parts per million carbon-dioxide equivalent, that is including other greenhouse gases, would give us a 66% chance to live up to the Paris 2° goal.[4] In other words, this threshold is weaker than Paris. US government websites, even two years into the Trump administration, report the current greenhouse gas concentration in the atmosphere. For 2018 it was 496 parts per million.[5] To be absolutely frank and honest, humanity has only negative emissions left.

If the Footprint tells us that we are already overloading the planet's biological capacity, one thing is clear: there will be no simple solutions. As with money, we can overdraw our account for a while—but accumulate debt in the process. When will our banker, the Earth, be no longer willing or able to extend the line of credit?

The Footprint's strength is its ability to give us an orientation, mainly because it works with observable data only. Ecological accounting is the basis of its method and its core purpose. That includes identifying the limits of the natural systems and providing numbers as supporting evidence. As Simon Upton, New Zealand's former Minister for Science and for the Environment and former Head of the OECD's Environmental Directorate, put it in his talks about the Footprint: "We know only one thing for sure about Footprint: Its results are wrong...but they are the best we have so far. Which is much better than having none. If you want to have better ones, get involved; help to develop more reliable data. Help to improve the method. Or produce something better yourself."[6]

Of course, the Footprint cannot answer every question. Identifying ecological viability is not a sufficient condition for a sustainable future. But it is a necessary one. To tell us what is: the Footprint does so currently best at the level of nation-states. The quality of data is crucial, such as the comprehensive trade statistics we have for countries. To come up with a country's Footprint, we can add up the imports and subtract the exports. These figures allow tracking the exchange of biological capacity.

Evaluating the Validity
of Ecological Footprint Results

In the academic, public policy, and popular literature, there are plenty of criticisms that challenge the Ecological Footprint method or its results. A good summary of one that is still valid and partially addressed is available on the Web.[7]

Question Sequence to Test Ecological Footprint

This sequence of questions enables you to find out whether the Ecological Footprint is fit for your needs as a sustainability indicator.

1. Does sustainability require that humanity's demand on nature stay within what the planet can renew?

2. If so, is it the case that biological regeneration is the limiting material factor for our economy? More specifically:

   a. Is it so that fossil fuels are far more limited by how much extra carbon ecosystems can absorb, rather than by how much fossil fuel is left underground?

   b. Given the water-energy-food-biodiversity-climate etc. nexus, are water, climate, soil, etc. input factors for regeneration, with regeneration being the outcome?

   c. Is it accurate to say that human demands on nature's regeneration largely capture, directly or indirectly, people's competing uses of nature? This means that non-competing uses of nature would not be counted as additional demand, since they are not materially limiting. Here an example of a non-competing use, photovoltaic (PV) panels, put on a non-productive area provide electricity without compromising biocapacity.

3. If so, is it reasonable to measure this planet's biological regeneration (or biocapacity) by tracking and recognizing the relative productivity of all the biologically productive areas available on Earth?

4. If so, is it also reasonable to assign to areas of the planet (such as regions, countries, farms) fractions of the total global regeneration? (i.e., which percentage of total planetary regeneration is harbored by each particular area?) Global hectares represent equal fractions of the planet's biocapacity.

5. To contrast what's available to what is being used by people, is it reasonable to map somebody's consumption on the mutually exclusive biologically productive areas needed to provide for this consumption? (The area is also expressed in global hectares.) Mutually exclusive means there is no double counting. The accounting only includes areas whose use excludes other uses.

6. Then, can we compare these two amounts: amount demanded (Footprint) versus amount available (biocapacity)?

7. Are Ecological Footprint accounts attempting to provide such an assessment?

8. Are there other measures available that answer this question of human demand against biological regeneration more accurately? (If this is the case, then use those!)

9. If no better ones are available, are Ecological Footprint accounts so far off that we would be better off without their results? 🌿

Ecological Footprint accounting provides a clear picture at the national level, both of supply (the biological capacity of the respective country) and of demand (the population's Footprint). The information is uniformly expressed in global hectares, the Footprint's core measurement unit.

One more detail about the method:[8] a country's National Footprint and Biocapacity Accounts depend on the size of its population, the quantity of an average person's consumption, and the intensity of resources used for the production of goods and services, addressing, among other things, how clean the respective technologies are.

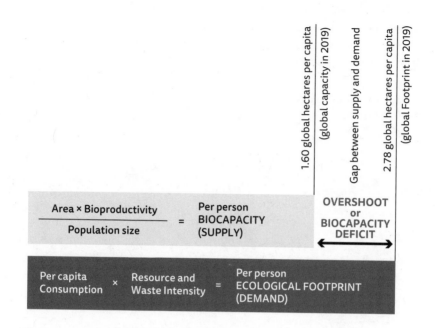

Figure 4.1. The difference between the global Footprint and the global bio-capacity is global overshoot. It can be expressed per person or in total. At the country level, the difference is its biocapacity deficit, if the country's Footprint is larger than its biocapacity. It is not necessarily overshoot, since the missing amount can come from either local overshoot or from used (or overused) biocapacity located somewhere else.

With all accounts done, we can have annual comparisons: How big is a country's Footprint? How much natural capital is available on the other side of the ledger? Residents of countries with a *biocapacity deficit* take in aggregate more from nature than the country's ecosystems can regenerate. Conversely, countries with an *ecological reserve* are countries that have more ecological capacity than their residents use.

Among highest-income countries, Australia, Sweden, Finland, New Zealand, and Canada are still in the black. Large portions of South America and many African countries are also among the creditors. But hardly any country along the Mediterranean, and neither India or China, are in the black.

Figure 4.2. Footprint compared to country's biocapacity, 1961 and 2016. The maps show biocapacity deficit (darker) or biocapacity reserve (brighter) of countries. For more detailed maps, and to see changes over time, visit data.footprintnetwork.org.

■ Biocapacity Deficit: Country's Footprint larger than its biocapacity
■ Biocapacity Reserve: Country's biocapacity larger than its Footprint
□ Data incomplete

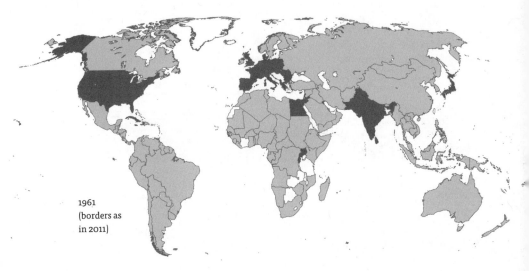

1961
(borders as
in 2011)

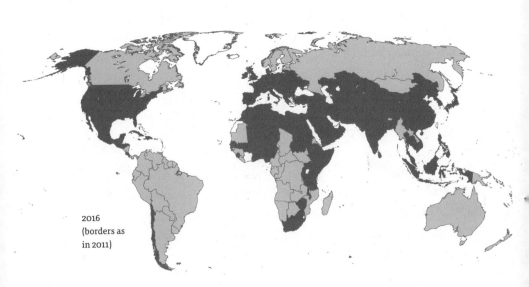

2016
(borders as
in 2011)

Countries with biocapacity deficits have, as discussed, three options for maintaining their level of consumption (see "biocapacity deficit or reserve" entry in Glossary for a brief summary). If they consume their biocapacity faster than can be regenerated, they may, depending on their financial means, import resources from other countries; they may also generate more waste than their area is able to absorb.

Carbon dioxide offers a good example. Every single day, industrialized nations pump enormous amounts of it into the atmosphere. Not taking into account wastewater and earthworks, carbon dioxide emissions make up about 80% (in terms of weight) of the industrialized nations' waste.[9] Even a country as large as the United States requires twice as much biocapacity to support its consumption than it has in its enormous territory. On the other hand, in several African countries (for example, Tanzania or Malawi), carbon dioxide's share is barely visible in their Footprint diagram.

Countries blessed with natural capital—such as Brazil, New Zealand, or several African countries—are not automatically among the winners. To be a winner, they need to treat their ecosystems with care. In addition, their population needs to benefit from the natural wealth of their country, which is not always the case.[10] Yet current market distortions disfavor natural capital, providing only a tiny portion of the benefits created by a value chain back to the those who take care of the natural capital.

But these biological resources are becoming more important: In the early 1960s, more than 60% of the global population still lived in countries that showed an ecological reserve, while today, a mere 14% do.[11]

If we calculate the sum of all National Footprint and Biocapacity Accounts, all based on UN statistics, we arrive at revealing numbers. More than 12.2 billion hectares comprise all of biologically productive land or water on our planet.[12] But in 2019, our demand amounted to 21.4 billion hectares. The difference is roughly 75 percent of the planet's biocapacity. Earth, hence, is in a state of overshoot.

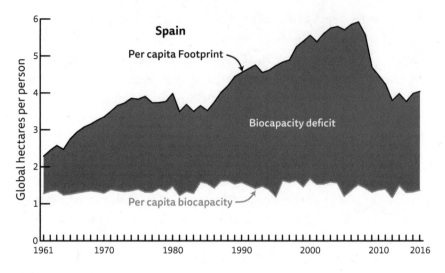

Figure 4.3. Spain: Ecological Footprint and biocapacity per person in global hectares per person. The rapid growth of Spain's Footprint corresponds to the fast-paced expansion of its infrastructure, demanding resources for its initial construction as well as its subsequent operation. The financial shock in 2008 dramatically reversed the Footprint—and made part of that infrastructure useless because its operation was no longer viable. Credit: Global Footprint Network—National Footprint and Biocapacity Accounts 2019 edition, data.footprintnetwork.org.

Meanwhile, the shortfall keeps growing. By now the global ecological debt (the accumulation of biocapacity deficits since the 1970s when global overshoot began) is the equivalent of over seventeen planet Earth years. Put differently, the planet would require seventeen years to pay off that mountain of debt, provided all processes, including the destructive ones, could simply be reversed and people would not harvest anything during that period.

Let's take Spain, a success story in the European Union. In the past 40 years, the country's population increased by about 10%. During that same period, the economy has grown markedly faster and expanded its physical infrastructure, which of course required more resources and more energy. Spain had enough income to import resources, and emitted plenty of carbon dioxide into the atmosphere (instead of absorbing it within its own ecosystems). It did not pay for

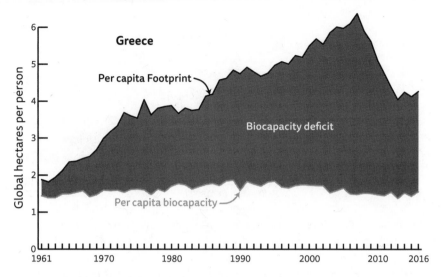

Figure 4.4. Greece: Ecological Footprint and biocapacity per person in global hectares per person. The rapid increase of its Footprint and Footprint's even more rapid reduction during its financial crisis resembles the situation in Spain. However, Greece's biocapacity deficit—as well as its financial one— is even worse than Spain's. Credit: Global Footprint Network—National Footprint and Biocapacity Accounts 2019 edition, data.footprintnetwork.org.

these emissions; but it did pay for its fossil energy. Then the financial crisis tremendously weakened Spain's economy. The future value of the already built infrastructure crumbled, and with it, economic activities. With less income, consumption shrank, as Figure 4.3 shows. Today Spain's Footprint is only twice as big as its biocapacity.

Both in Spain and in Greece, the cost of buying resources for rapidly expanding infrastructure burdened the economy. As a result, both countries' Footprints markedly shrank, which pleased neither the Greeks nor the Spaniards.

An even more extreme example is North Korea. After the collapse of the Soviet Union, the country no longer received oil or coal from the Soviets. In addition, the Soviet Union's collapse challenged China economically. Further, to North Korea's detriment, China's competition with the Soviet Union for influence and favored status in North Korea eased up. As a result, China reduced its shipments of rice to North Korea.

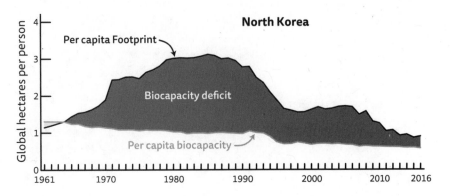

Figure 4.5. North Korea: Ecological Footprint and biocapacity per person in global hectares per person. North Korea has yet to recover from the rapid and significant Footprint contraction caused by the disintegration of the Soviet Union. Its biocapacity per person keeps diminishing. Credit: Global Footprint Network—National Footprint and Biocapacity Accounts 2019 edition, data. footprintnetwork.org.

Less oil and coal from the Soviet Union caused North Korea's biocapacity to decrease: with less available fertilizer and not enough fuel for their tractors, the North Koreans could no longer produce sufficient food—in addition, they received less rice from China. As a consequence, people starved. More than one million people are feared to have died from hunger as a fallout. The situation was made worse by the country's ineffective and rigid administration. It meant that there was no escape from local ecological limits. Without other sources for additional imports and without sufficient local resources, there simply was not enough to sustain the people.[13]

North Korea did not trade with other countries by choice. Our planet as a whole does not trade either with other planets, limited by the physics of interplanetary trade rather than ideological choices. North Korea is a cruel example of how resource scarcity directly affects people's quality of life.

Resource dynamics are also closely linked to the Arab Spring which has extended into a rather chilly fall. In Arab Spring countries, a rapidly growing population was faced by resource constraints and hence tightening economic opportunities. Egypt's economy has a

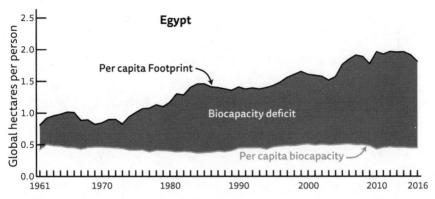

Figure 4.6. Egypt: Ecological Footprint and biocapacity per person in global hectares per person. The growing deficit in biocapacity is causing growing problems for Egypt's economy. Credit: Global Footprint Network—National Footprint and Biocapacity Accounts 2019 edition, data.footprintnetwork.org.

massive focus on boosting fossil fuel extraction. Still, domestic oil demand consumes the bulk of its oil extraction; also, its electric demand often outdoes its production capacity leading to regular blackouts during peak demand, leading to popular protest movements. The food constraints are also severe: by now, 60% of the agricultural products Egyptians consume are imported. Growing scarcity curtails opportunities and accelerates social conflicts. In 2006, before its civil war broke out, Syria experienced a drought that reduced people's incomes particularly in rural areas and lead many to migrate to urban areas. Many researchers consider that the drought triggered the civil war.[14]

Overexploitation of ecosystems is by no means a new phenomenon. The earliest evidence of the destruction of an ecosystem dates back to the Sumerians, starting in 2400 BCE.[15] Because of the geological conditions in the valley between Euphrates and Tigris, food production in that area was always difficult. In the spring, both rivers carry huge amounts of water from the snowmelt, yet between August and October, they are mere streams with minimum flows. But summer and autumn are exactly the seasons when agriculture badly needs water. The Sumerians were smart and developed techniques to store water and use it for grain cultivation—one of the earliest

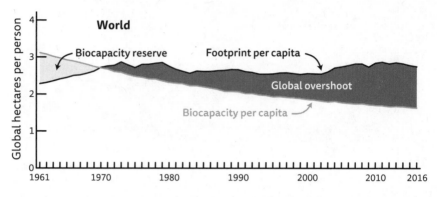

Figure 4.7. Humanity's Ecological Footprint and Earth's biocapacity in global hectares per person. With the beginning of the 1970s, the human enterprise entered the state of overshoot. Credit: Global Footprint Network—National Footprint and Biocapacity Accounts 2019 edition, data.footprintnetwork.org.

irrigation systems in the world. The productivity of their ecosystems increased, as did their wheat yield. It was thanks to these developments that one of the earliest advanced civilizations was able to emerge.

Summers in those latitudes are very hot, often over 100°F. Water in ditches and on cropland evaporates quickly. Left behind is the salt that was originally dissolved in the water. From the year 2000 BCE on, increasingly frequent records reported that "the earth turned white." The consequence was a serious decline in wheat production, caused by the increased soil salinity. Salinization has remained, by the way, one of the main problems of irrigation to this day. Everywhere.

This early example from the Sumerians demonstrates the characteristics of overshoot.[16] First, growth occurs and developments accelerate (irrigation increased the productivity of the Sumerians' ecosystem). Second, a threshold is crossed, with a significant disruption of the system (from a certain degree of salinization onward, plants reacted negatively and yields dropped). Third, the process is accompanied by a lack of people's awareness or by delayed feedback so that the mistake can no longer be corrected. Put differently, people

learn too late (the Sumerians stood no chance: the process of salinization had not yet been understood).

Generally, overshoot is not intentional. To the participants, it initially seems merely an undesirable side effect. And it tends to creep
up on them. That is what makes overshoot so dangerous.

The fate of the Sumerians, their unintentional mismanagement
and overexploitation of their ecosystems, has repeated itself countless times in other places and different cultures. Today, though,
the extent of the problem is far more worrisome. We are no longer
dealing with exclusively regional forms of ecosystem degradation;
instead, overconsumption has expanded to the entire planet. The climate problem, a consequence of decades of worldwide overexploitation, may well be its most obvious sign.

National Footprint and Biocapacity Accounts can show when humanity's global Footprint exceeded Earth's biocapacity: at the beginning of the 1970s.[17] That this historic event occurred during a period
of previously unknown material growth is surely no coincidence.[18]

If we look at the situation through the lens of the annual biocapacity assessments, we arrive at a symbolic date: Earth Overshoot Day.[19]
In 2019, this day was July 29. From January 1 to July 29, humanity had
already demanded as much from nature as Earth was able to renew
during the entire year of 2019, from food to energy and construction
materials; in addition, Earth still had to absorb our solid, liquid, and
gaseous wastes. From July 30 forward to the end of that year, humanity lived on its ecological credit card. The ecosystems' natural sinks
are getting filled, their stocks reduced.[20]

Overshooting can work for a while. But if we continue along this
path, a widening gap opens between incomes and expenditures. And
even if it did not widen, it cannot be maintained for long. On the contrary. Since the 1970s, Earth Overshoot Day has generally been moving forward in the calendar. In 1990, it was October 11; in 2000, it was
September 23; and in 2010, August 7.

Biocapacity deficits which accumulate to ever larger ecological
debts are a risk to cities, regions, and countries. But in contrast to

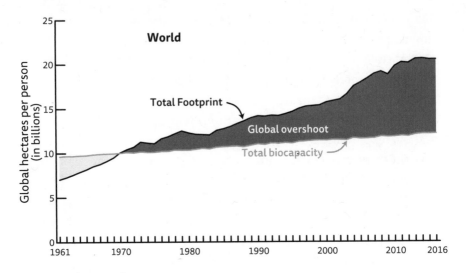

Figure 4.8. Humanity's Ecological Footprint and Earth's biocapacity in total. This shows the same trend as Figure 4.7, just in totals rather than per capita, with Footprint and biocapacity crossing over in the same year. It also shows that total biocapacity has kept increasing, mainly due to the intensification of agriculture. Whether this increase can be sustained in the long run is not clear. Humanity's Footprint and Earth's biocapacity are expressed in billions of global hectares. Credit: Global Footprint Network—National Footprint and Biocapacity Accounts 2019 edition, data.footprintnetwork.org

financial capital, which—in its different forms, such as cash, shares, or bonds—can be compared and exchanged, natural capital cannot be simply moved or converted.

Instead, different usages of a resource may even compete with each other. If Brazilian rain forest is cut down to make room for sugar cane plantations to produce agrofuel, fewer trees are left to absorb carbon dioxide. If fishing grounds collapse in Canada, the ecological pressure grows on grazing land to produce animal protein on land. Also, human Footprints compete with the Footprints of wild species for the planet's biocapacity.

During the course of its history, humanity has entered ever deeper into nature, has colonized and shaped it to suit human purposes. Currently, the crucial losses of natural territory are taking place in the tropics and subtropics. In South America and the Congo Basin, unimaginably vast areas of forest are being cut. In Brazil, for example, along the Amazon more than one million hectares of tropical forest are lost every single year—ten years ago, it was even more than three million hectares per year.[21] If the Brazilian government elected in 2018 on a platform advocating more aggressive exploitation acts on its promises, the slowed down deforestation may accelerate again. The general rule is this: the greater the global Footprint, the more land is claimed by humanity, and the less is left for wild animals and plants. And with that, biodiversity suffers. So the question arises, how far can, or how far should, humanity go?

According to the Living Plant Index, the average population size of vertebrate species in the wild has decreased by 60% over the past 40 years.[22] Even the modest goal of just slowing down the loss of biodiversity looks rather unrealistic these days. Footprint accounting does not keep a ledger of the numbers of animal and plant species. Our National Footprint and Biocapacity Accounts, limited by UN data, do not even distinguish whether forest is mixed or monoculture. But by comparing the quantity of human demand against available biocapacity, Footprint accounting does give an indication whether there is sufficient biocapacity available for wild animal and plant species.

More than twenty years ago, the international discussion of biodiversity prevailed in its push to reach agreement on a practical goal: placing under protection roughly 10% of each biotope on the Earth's surface. The goal has been reached and far extended by the Aichi Biodiversity Targets of 2010,[23] but even those more ambitious targets have proven insufficient in stopping the drop in the number of species. Most likely, we will continue to lose enchanting birds, primates, or even the rhinoceros. Nevertheless, humans, one of the most flexible and widest-spread species on this planet, will survive—in an impoverished biotope with artificial, intensely used cultural landscapes. Not a great prospect, but not an unlikely one, either.

If we really wanted to stop the mass extinction happening right now, we would have to return portions of the areas used by humans to wild animal and plant species. The size of any of these areas would not be the only factor, but valuable regions would require special protection. The geography of biodiversity is well understood both globally and regionally.[24] "Hot spots" that would have to be given priority are in Central America, Western Amazonia, the Cape of South Africa, and in the mountains and plains of eastern Africa. Others can be found along the coasts and on the islands of the Mediterranean regions, in southwestern China and the bordering areas from Burma to Vietnam, as well as in Indonesia and New Guinea. Large parts of Madagascar are among these valuable areas, as are numerous islands in the Pacific and the Indian Oceans. It is a complicated challenge in light of a growing global human population and its hunger for area. The situation is not entirely hopeless, though; quite often biotopes with rich biodiversity have developed in places no one would have ever expected: above all, in the ecological niches of large cities.[25]

The biggest challenge humanity faces at the beginning of the 21st century is this: How can we, together with the other species, lead rich and fulfilled lives within the constraints of this one planet's biocapacity?

Like any indicator, the Footprint follows a particular logic. It accounts for some things more successfully than for others. For example, it only indirectly gets at water scarcity, biodiversity, or damage caused by environmental toxins. And yet, it is the most comprehensive account that currently exists to compare overall human demand against planetary regeneration, embracing, yet going beyond, the climate change dimension. It is also readily available, over some historical time, for more than 150 countries. And the Footprint not only identifies the ecological constraints of the planet and all its ecosystems but is able to quantify them in ways people can understand.

While our planet is finite, human possibilities are not. The transformation to a sustainable, carbon-neutral world will succeed if we apply humanity's greatest strengths: foresight and innovation. The good news is that this transformation is not only technologically possible, it is also economically beneficial and our best chance for a prosperous future.

Mathematically, four factors determine the extent of our over-shoot. The first two determine the Footprint side, the other two the biocapacity side.

1. Reducing the Global Population

Global population growth can be reduced and eventually reversed toward a shrinking human population. We owe it to our children. The ecological benefits accrue slowly, but over time accumulate and become vast. But the social benefits from encouraging smaller families emerge fast, with tangible benefits to living children's prospects. Encouraging smaller families means ensuring that women have as many rights and opportunities as men had for the last decades, if not centuries. There is really no downside to that.

Reducing human population applies equally to industrialized and rural regions, and everywhere else in the world. It means supporting women's educational and employment opportunities, warranting equal rights, and securing access to reliable and safe family planning for men and women.[26] Even a few years of additional schooling and some microcredits produce numerous positive outcomes. As the number of children falls, the likelihood of an education and good health rises. The potential for violence is reduced. Societies with large numbers of young people—and especially young men—have been shown to be relatively prone to violent conflicts and wars.

If young women are kept away from public life or, as a financial insurance strategy, are predominantly married off to older men, many young men are left behind as well. Their frustration easily turns sour. Frustrated youths are also recruited for destructive causes. In the context of development policies, interpreting investments in women as merely a question of gender politics falls short of illuminating real potential. In lowest-income countries, focusing on turning demographic trends while enabling opportunities by accelerating zero-carbon energy access everywhere and boosting agricultural productivities among small holders may be among the most promising core strategies for boosting development that generates lasting outcomes. In short, investments in women, and their ability to fully participate, benefit society as a whole.[27]

2. Reducing the Footprint per Person

The opportunities for securing, or better even enhancing, people's lives on a smaller Ecological Footprint are enormous and go beyond finding more happiness consuming less. Big scalable gains come from offering infrastructure that enables more effective ways of living.

Three domains are particularly central to this theme:

How We Design and Manage Cities

The way our cities are shaped determines both heating and cooling needs as well as transportation. Cities are shifting rapidly. Eighty percent of an even larger world population is expected to live in cities by 2050. This translates into a near doubling of urban populations by then. Consequently, city planning and urban development strategies are instrumental to balancing the supply of natural capital and population's demand. Mobility needs and energy efficiency of housing shape cities' long-term resource dependence.

Energy—How We Power Ourselves

Carbon emissions currently make up the biggest share of humanity's Footprint. Decarbonizing the economy is our best possible chance to address climate change, and would help realign our Ecological Footprint and the planet's biocapacity. Decarbonization is driven not only by reducing fossil, but by driving efficiency. In the past 40 years, technological advances have markedly increased resource efficiency in the production of goods and service, mostly for energy. As a result, the average *per capita* Footprint has remained relatively flat in high-income countries. Of course, an increase in resource efficiency—for example, through more efficient cars or better insulated homes—does not increase the available biocapacity. Increased resource efficiency simply means that we can gain more goods or services from a given quantity of resources.

Companies do react to political signals that tell them to increase their resource efficiency; however, for this to provide ecological benefits, clear and long-term measures are needed. Consumers have the power to exercise pressure in this regard. An efficient technology can ultimately consume even more resources, when consumption rises faster than the technology's reduction of the environmental impact per unit. This boomerang is called the *rebound effect*. Countering it requires not just technological but political measures. For example, eco-taxes on energy will incentivize energy efficiency, while also funneling off the extra financial gains to support societal goals, thereby preventing the rebound effect.[28]

Food—How We Produce, Distribute, and Consume It
How we meet this basic need is a powerful way to influence sustainability. Avoiding food waste, using sustainable agriculture, and eating lower on the food chain lowers the Ecological Footprint. Currently food production uses over half of our planet's biocapacity.

3. Restoring and Nurturing Productive Areas

Biocapacity can also be shifted, although not as much as the Footprint. Through careful measures, degraded lands such as semideserts or salty soils can be brought back for cultivation. Terrace farming has historically been successful, and irrigation can indeed increase the productivity of soils, although the gains are often temporary. Above all, wise land management is necessary to ensure that biologically productive lands do not degrade or shrink, for example, as a result of urbanization, salinization of soils, or expansion of deserts. Also care needs to be given to biodiversity, as excessive use reduces opportunities for wild species.

4. Increasing Productivity per Hectare

An increase in productivity per hectare depends on the nature of the respective ecosystem and on its management. Agrotechnology can increase productivity and at the same time diminish biodiversity. Through energy-intensive agriculture, which may also rely heavily on pesticides, yields can indeed be increased. But this increase will come at a cost: a greater Footprint caused by these inputs. Over time, the soil may become depleted, and yields drop long term.

However, the biocapacity of soils can be maintained or even strengthened through a number of measures, such as protecting soil from erosion and other causes of degradation. Rivers, wetland, and watersheds need protection to stabilize the water supply and to maintain healthy forests, croplands, and fishing grounds. Finally, proactive land management will strive to mitigate the effects of climate change in order to stabilize harvests. If done carefully, sustainable intensification allows for concentrating agricultural production, and leaving more space for wild species.

Thriving lives within the means of our planet are not out of reach. Plenty of solutions exist in these four major areas for improving the chances of a sustainable future. All these areas are characterized by enormous inertia: they cannot be shifted rapidly. This means we can either lock ourselves into highly valuable assets or, equally possible, lock ourselves into infrastructure traps that severely limit economic possibilities in the long run.

CHAPTER 5

FOOTPRINT AS COMPASS

How Much Biocapacity Do We Need for a Good Life?

People with a lot of money can afford a lot of nature. Purchasing power and the Footprint are closely linked. But both are unevenly distributed on the planet. While the industrialized countries consume the lion's share of natural resources, newly advanced economies such as China or India are in an unparalleled race to catch up. Yet a less energy- and resource-intensive economy and lifestyle are possible without losing quality of life. Meanwhile, though, many low-income countries are falling behind. This is not a simple situation.

Global commuters are the nomads of our time. They live in the halls, waiting areas, and shopping malls of international airports. Their cell phones and laptops are their eyes and ears.

- A worker with family in Scotland and a job on an oil rig off the Norwegian coast commutes every week between Glasgow and Bergen. On the plane, he encounters familiar faces. He owns two cell phones and two wallets, one for each country.
- An English television presenter is sick and tired of riding the crowded London Tube. She now lives in Barcelona, takes whenever possible the low-cost airline, and enjoys a cheaper and more pleasant life in Spain.
- A lawyer specializing in international law lives in San Francisco. Most of his clients are in Asia, especially in Japan and South Korea but increasingly also in China. If he doesn't fly there every week or two, he will stop getting contracts.

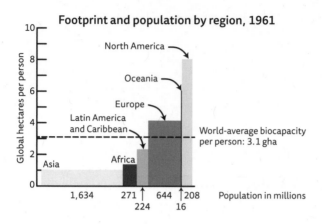

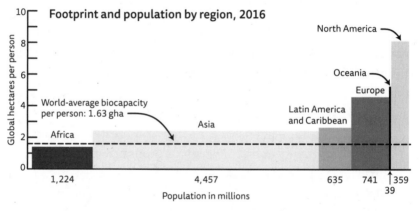

Figure 5.1. Distribution of per person Footprint and population by region, in 1961 and 2016. The areas of the bars represent the total Footprint of each region. Credit: National Footprint and Biocapacity Accounts, 2019 edition.

On the internet, global commuters share their experiences. They chat about stress or no stress, about the impact of the fast life on family and relationships, about work contracts and insurance policies, about strategies to avoid expensive flights during rush hour periods, and about the things needed for daily life.

Someone commuting every week will take four return flights a month, which easily adds up to 100 flights per year—and a whole lot of jet fuel. The energy Footprint of a global commuter surpasses the average by at least a factor of ten. And so continues a basic historical pattern of industrialized societies, increase of energy consumption.

As a general rule, the higher the income, the higher the Footprint,

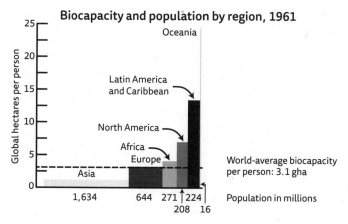

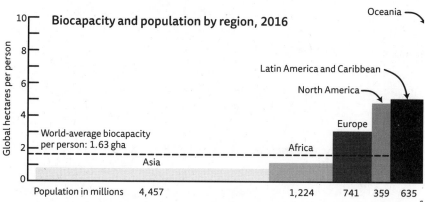

Figure 5.2. Distribution of per person biocapacity and population by region, in 1961 and 2016. The areas of the bars represent the total biocapacity of each region. Credit: National Footprint and Biocapacity Accounts, 2019 edition.

and the higher still the relative share of energy consumption (or carbon Footprint) in that Footprint. In industrialized urban societies, the carbon Footprint makes up over half an individual's Footprint, while the relative share contributed by food is falling.[1]

Ultimately, the big question is not the Footprint—but people's well-being. The fact that there is only so much biocapacity is a side condition. The question is how we can generate best possible lives, given the one-planet context we face. This very challenge is at the core of sustainable development, a quest that became a global venture during the run-up to the 1992 Rio Conference,[2] and has become ever more central to national and international agendas ever since.

The Search for Sustainable Development

How can we know whether a country moves toward sustainable development? The essence of *sustainable development* is quite straightforward. Development is shorthand for committing to well-being for all. Sustainable implies that such development must come without depleting the future. In other words, development is required to occur within what the planet's ecosystems are able to replenish season after season, year after year. It means we need to operate within the means of our one planet.

Kate Raworth calls this double condition the "safe and just operating space for humanity." She depicts it with two simple circles (the doughnut—see Figure 5.3), the inner circle representing minimum social conditions, while the outer circle stands for maximum ecological demand on the planet.[3]

We can measure this "outer doughnut" condition using the Footprint, as an approximation for whether we operate within the means of our planet. To evaluate to what extent development fits within the planetary constraints, we can compare people's Footprints against how much biocapacity is available per person.

Human well-being, or the "inner doughnut" condition, can be measured in a number of ways, but possibly the most prominent metric is the United Nations Development Programme (UNDP)'s Human Development Index (HDI).[4] This outcome measure evaluates a country's development achievements by combining three scores: one for people's life expectancy, one for their literacy levels, and the third one for the *per capita* gross national income of the place they live. Each of the three get scaled from 0 (worst) to 1 (best), and then the three components are averaged out. UNDP considers an HDI value of more than 0.7 "high human development"; 0.8 is "very high."

To depict these two dimensions jointly in quantitative terms, we plot them on a two-dimensional graph (two axes). The doughnut, while simpler and easier to understand, only offers one dimension (one axis from inside to outside). To illustrate performance against *global* sustainable development, we can track these two dimensions against each other. Figure 5.4 shows a diagram where the two measures are combined.

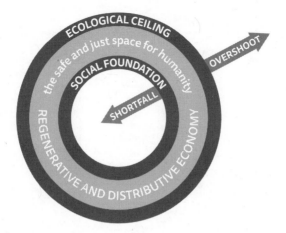

Figure 5.3. The healthy doughnut. Kate Raworth defines the doughnut economy as one that meets the needs of all within the means of the planet. She calls the doughnut of social and planetary boundaries "a playfully serious approach to framing that challenge, and it acts as a compass for human progress this century." Credit: after Kate Raworth (Doughnut Economy).

On the right side of the chart,[5] we see two dotted vertical lines, the thresholds for the HDI. To the right of the threshold of 0.7 for "high" and the one of 0.8 for "very high development," we find many countries in Europe and North America, several in the Asia-Pacific region, and some in South America.

The horizontal dotted line in Figure 5.4 shows how much biocapacity is available per person (1.63 global hectares of biologically productive area in 2016, or one Earth equivalent.). This line is essentially the "one planet line" or the biocapacity threshold. E. O. Wilson would argue that we should aim for the half planet line in order to leave space for wild species. The quadrant on the top right, labeled "global sustainable development quadrant" is where the world average would need to be if we want to have sustainable development. Actually, for true sustainable development, it is only half of the darker part: that one defined by very high human development, and half the planet's biocapacity. This quadrant corresponds to the grey ring of Kate Raworth's doughnut (see Figure 5.3): the safe and just operating space. It shows the place of high human well-being, in a way that is physically replicable across the world.

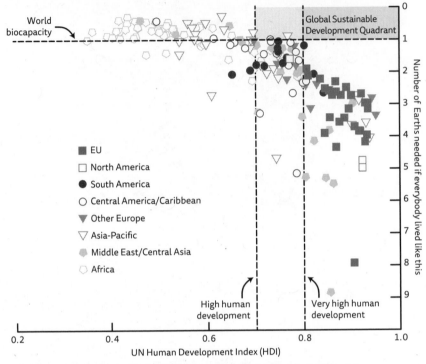

Figure 5.4. The Global Sustainable Development Quadrant: Tracking countries' sustainable development performance with the United Nations Human Development Index (UNDP's HDI) and per person Ecological Footprint, by regions. The Footprint is expressed in Earth equivalents, meaning how many Earths would be used if all humanity had that Footprint. In 2016, a 1.63 global hectare Footprint corresponded to one Earth equivalent. Lower positions in the graph means that more Earths would be needed. Ecological Footprint results are from the 2019 edition of the National Footprint and Biocapacity Accounts; HDI values are from UNDP's 2018 Human Development Report. All values are for data year 2016.

Most of the countries with high human development also feature a Footprint that exceeds the one-planet line, let alone the half-planet line. In 2016, only three countries meet these two criteria simultaneously: Jamaica, Philippines, and Sri Lanka. Cuba, the Dominican Republic, Ecuador, and Uruguay met the human development criteria, but also with a Footprint that was slightly above what's globally replicable. Uruguay even reached "very high" human development. No country meets E. O. Wilson's vision: very high human development within the means of half-Earth.

None of this means that Jamaicans, Filipinos, or Sri Lankans are happier than the residents of other countries or that life in this quadrant is automatically better, particularly in the short run. Having a high quality of life is of course easier with more resources. The same way as things are easier to achieve if we have more financial budget. But the point here is that our global resource budget is constrained; therefore, how can people have the best life within our ecological opportunities? As the chart shows, these three countries manage with relatively few resources to guarantee high life expectancy, high literacy levels, and medium-level incomes.

Over the past decades, many countries, including China and India, have shown notable improvements in their standard of living as measured by the Human Development Index. India with a *per capita* Footprint of 1.2 global hectares still remains significantly below the one-planet threshold of 1.63 global hectares *per capita*. Its HDI reached 0.64 in 2017. China's Footprint, meanwhile, has surpassed the one-planet threshold by more than factor 2. But it has reached a high human development of 0.75 HDI in 2017. According to UNDP, it crossed the high-development threshold of 0.7 HDI around 2010.[6]

Living standards in Africa, as measured by the HDI, have also somewhat improved, though not in many countries. Still only four, possibly five, countries on the African continent have reached high human development: Algeria, Botswana, Gabon, and Tunisia. The fifth is Libya, but given current trends, it is unlikely that its HDI is still above 0.7.

This HDI-Footprint diagram provides two insights:
1. There seems to be a historical pattern that higher human development comes handin hand with higher Footprints.
2. The pattern is not a hard line nor a law of physics—countries deviate widely from the average trend.

The second insight is powerful as it allows us to identify different ways the pattern can be broken. Even more interesting is to follow the trajectories of countries over time on this diagram.[7] It seems that also countries' time trajectories, by and large, follow the pattern painted by countries in the HDI-Footprint diagram.

If we are truly committed to sustainable development, this HDI-Footprint diagram serves as an outcome measure. Ultimately, the Sustainable Development Goals (SDGs) have to move the world average into the global sustainable development quadrant. SDGs provide many strategies. Do these goals add up to the outcomes that are ultimately needed: better lives for all within the resource budget of our one planet? Currently, it looks like SDGs are not sufficiently strong to drive such positive country trajectories.[8]

Mathis used to say that if he were president of the World Bank, head of a development agency, or running the SDGs, the first thing he would do is to dedicate one of his office walls to a large printout of this diagram. Then he would invite all project and program leaders to ask them for evidence they have that they have been moving populations toward the quadrant. Even how much movement they have been able to generate per dollar investment. If those leaders cannot make the case, he would ask them to reform their strategies. And if they could not reform, he would suggest them to consider other employment opportunities. The bottom line: No project should be approved that is not helping to move humanity cost-effectively into the quadrant. Every other project is wasting precious resources and undermines humanity's ability to succeed.

For a video explanation of the diagram, you may like to watch Mathis Wackernagel's TEDx talk on YouTube.[9]

Is Sustainable Development Being Applied?

By now we can, admittedly, observe a slight decoupling of economic growth on the one hand and resource and energy consumption on the other hand,[10] but resource consumption in Europe—and essentially in all industrialized countries—remains by no means sustainable. It cannot be maintained over a longer period of time.

The economy suffers from obesity. It continuously devours oil, coal, biomass, metals, and minerals. But instead of using them efficiently, it spews out large amounts of this flow of material more or less undigested.

Europe (EU-28) comprises about 7.2% of the global population, but their Ecological Footprint claims almost 20 percent of the planet's biocapacity. The demand of the EU-28 is roughly about twice the size of its supply.[11] One of the fundamental problems: all across Europe, the Footprint remains high even though it has notably fallen in agriculture. The Footprint of an average Swiss, for example, is 4.7 global hectares and thus slightly above the European average. But with all the mountains and their large population, the biocapacity in Switzerland per person is barely ⅔ of what's available per person in the world. Europeans are taking a mid-field position in the league of industrialized states. At 7.7 global hectares, the Footprint of a citizen of the United States is considerably greater. Anyone who has ever visited the United States knows why: everything is bigger and more spread out. Just think of North American settlement structures with their never-ending suburbs and detached houses, mile after mile after mile. This "drive-in utopia" originated in the dream of life in the country, admittedly with a highway on-ramp leading to the city. It all comes at a price, not only of dollars but also of Footprint.

Let's not neglect the question of meat consumption—just from a resource perspective. For every kilo an animal puts on, it has to eat several times as much in feed. By the way, the live weight of all domestic cattle alone—not considering that of pigs, sheep, goats, or salmon—is two or three times the collective weight of the 7.7 billion people on the planet.[12]

Typically, each person's Footprint depends to a large degree on the home they live in, how much energy they use to heat or cool it, what mobility habits they have, how they get to work, and how much they travel. What they eat matters, too. In the end, it hardly makes a difference whether or not they recycle their yogurt containers. (Not that you should not, just that it does not shift the sustainability needle sufficiently.)

The Footprint does not ask people to abstain from consumption; it does not advocate suffering and sacrifice, it is not an ecologically disguised sermon, and the last thing it aims to do is make our lives unpleasant. It is just a metric. If anything, it is designed to make people's lives better: It is about enabling us to lead good and rich lives given the context of our one planet. But in order to do so, we need new ideas about what a good life looks like and how we can shape it to our satisfaction.

The Footprint as an accounting system shows clear patterns of individual resource consumption. That average Europeans in their daily lives require about half the biocapacity of Americans is also due to the fact that Siena, for example, has a greater population density than Houston and hence less traffic and more compact houses. Whether we look at city planning, architecture, traffic systems, industrial production, or agriculture—the potential for saving resources in high-income societies is so vast that a flourishing life on a smaller Footprint is absolutely possible. It is not only possible but will become necessary.

If it is so logical and simple, why are most countries and cities not working with such Footprint budgets? The reasons are simple: as long as sustainability does not pay off for the individual player and is not tied in with good economic incentives, it will hardly catch on. Footprint assessments in themselves do not directly address what fiscal or political steering tools might contribute to a sustainable economy. But it can serve as a compass for such tools. Year after year, it records where we are and where we are heading.

As an example, let's take a look at the Philippines. *Bat people* is what the poor are called who have found no homes even in the slums of the 20-million juggernaut that is Manila.[13] They live in shanties suspended below the many bridges of the Philippine capital. Above, cars, trucks, and buses thunder by on concrete roads, while directly below, an estimated 150,000 bat people spend their lives.

Their shanties are carefully put together out of bamboo slats and cardboard, with often nothing but rubber mats as floors. These box homes are a few meters long and wide. It is impossible to stand in them, so people have to crawl inside their shacks. In most cases, these shacks are home for an entire family. In one corner of the floor is a hole that serves in turn as kitchen drain, garbage dump, and toilet.

The dark, oily water a few meters below is almost stagnant. In this brackish river, children splash and paddle for fun and to cool off somewhat in the tropic temperatures. The filthy water tends to cause inflammation of their eyes and ears as well as other illnesses. There's no money for medication.

The bat people mostly survive on small jobs, unloading boats in the harbor, doing construction work, or driving jeepneys, Manila's rickety cabs. Because of their jobs, they occasionally pass through the office districts with their glass towers and artificial waterfalls or ride past the enclaves of the rich, who live behind barbed wire in air-conditioned houses.

The Footprint of the bat people, too, can be calculated. Their children may occasionally run along the busy road that goes over their bridge to the nearest McDonald's. But because they have no money, they merely gaze. The construction material of their shanties is basically garbage, erected on what is not a building plot; they claim no area. The largest portion of their expenses by far, economically as well as ecologically, is spent on food. The bat people's personal Footprint lies far below the Philippine average—and that average itself is only 1.2 global hectares, $\frac{1}{7}$ of the American and $\frac{1}{4}$ of the European Footprint.

Manila is simultaneously poor and rich, like all metropolitan areas across the world, Africa and Asia included. Every year millions of rural residents are drawn to the cities, driven by necessity or motivated by hope. By 2050, the global population will have grown by another two or three billion people, and if we ask where most of them will live, the answer is already clear: in the megacities of Asia, Africa, and Latin America. The vast majority will subsist in slums with living conditions below any standard of human dignity, where running water, working toilets, and electricity remain distant dreams. Nevertheless, many will succeed in building modest lives for themselves and getting their children educated.

Today 60 percent of humanity lives in Asia. The average Footprint in this part of the world, excluding Russia, is 2.4 global hectares. In 2016, that average already exceeded the world's biocapacity of 1.63 global hectares, and, by even a larger margin, the 0.76 global hectares that existed per person in Asia, excluding Russia. It goes without saying that if that many people increased their demands even slightly, which seems desirable from a social perspective, the total resource effect is considerable.

China plays a special part in this dynamic.[14] In spite of its one-child policy, China's population has markedly grown in recent history (albeit significantly more slowly than in India, just as India's population has grown more slowly than Mexico's). But at the same time, China's Footprint has increased as well. The Footprint of residents of Shanghai, Beijing, or any provincial capital may nowadays be similar to the per-person Footprint of a European city. The Chinese carbon Footprint has increased more than tenfold since 1961, and its total Footprint now amounts to 3.6 global hectares, more than double the world's per person biocapacity. Yet, only half of China's population has been moved from a rural agricultural existence to an urbanized industrial one, a trend still in full swing. Impressively, in recent years, the Chinese Footprint has noticeably flattened off. The rapid increase in per person Ecological Footprint shown in Figure 5.5 captures this rapid transformation. As a result, China is now massively dependent on imports and on disposing of its waste outside its territory.

Explosive economic growth, a rapid increase of its carbon Footprint, and overall consumption of resources—these are true in both China and India. Together the two countries currently make up ⅖ of the global population and will soon top that number. Already they are economic heavyweights. And both strive to join the world's top economies over the next decades.

Here's a thought experiment: Imagine China were to live the American dream. Millions of Chinese, particularly in the booming regions in the country's south, already lead lives comparable to those of Europeans or Americans. They drive cars, eat more protein, take computers and cell phones for granted. They move into new homes in the suburbs. Let's imagine this lifestyle were practiced not only by a minority of the Chinese, as is the case today, but by all of them. If they all lived in 2050 the way Americans do today, they would consume so much more meat and beer that their consumption of wheat and meat would triple. If this were to happen, China alone would require almost the entire biocapacity of the planet.

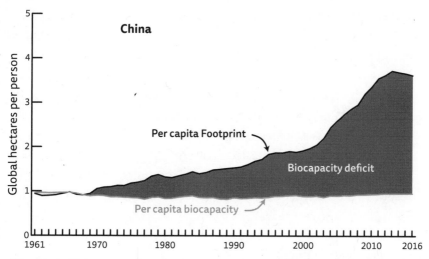

Figure 5.5. China: Ecological Footprint and biocapacity per capita in global hectares per person. China's per capita Footprint has doubled from 2000 to 2013 and then started to level off, in spite of rapid economic expansion. Credit: National Footprint and Biocapacity Accounts 2019.

Already Chinese consumption of steel—about 800 million tons annually—reflects almost ½ of the global demand, leading the way in *per capita* consumption worldwide. And if a Chinese citizen's current annual paper consumption of 50 kilograms were to jump to the American consumption level of about 300 kilograms, China would annually require twice the world's current production of paper. Finally, if China caught up with the United States in *per capita* oil consumption, considerably more than today's entire annual oil demand would be needed. Yet daily extraction volumes can hardly be raised much, in spite of fracking; already the output is reaching capacity limits. In short, if 1.4 billion Chinese were to try to increase their current Footprint of 3.6 global hectares per person (up from 1.9 in 2000) to the 8.3 global hectares *per capita* in the United States—it simply could not work. Yet the trends signal that in the coming fifteen years they will try to again double their resource consumption. Even adopting European levels would take them far beyond the physically possible.

If it is true for industrial nations that they won't be able to maintain their large Footprint over the long run, the situation for the whole community of BRICS countries is even more difficult. The way China, India, Brazil, and Indonesia right now race to also reach European or North American levels of consumption—namely by following the development models of those countries—cannot continue much longer. China's and India's impressive growth rates may fall sharply, and the resource-intensive economic boom could lead to allocation conflicts and ultimately resource wars. A genuine chance at development exists for countries with more recent urbanization and industrialization patterns only if they place their bets on energy- and resource-extensive economic models and lifestyles. Here, too, the Footprint can show the way.

A Case Study: Yemen

The residents of Sana'a, Yemen's capital, live in challenging times. Water has been a major theme: A few fountains still exist in their fairy-tale old town with its clay buildings, and old people remember drawing water from them. By the late 1960s, a new era began with the arrival of the first diesel water pumps in the country. Suddenly there was an abundance of water. However, the development has proven costly. The water table has since dropped from about minus 20 to 40 meters (which could still barely be pumped by hand) to minus 800 to 1,200 meters. And in the Sana'a valley, it keeps sinking by another 6 to 8 meters a year.[15]

Before the war, anyone in Sana'a with money has access to the public water supply. About twice a week, water rushed through their pipes. This is when residents knew to fill their homes' private tanks to tide them over the days without water. No one would know for certain when the water will return. But most residents in this city of two million inhabitants depended on water tankers, vehicles like those that in some other countries deliver heating fuel but that here are more rickety and more colorful. In Yemen, water, the most basic of all nutrients, is expensive. In contrast to its high-income neighbor Saudi Arabia, Yemen counts among the countries with the world's lowest *per capita* income.

At 2,200 meters above sea level, Sana'a has a relatively high altitude. It is surrounded by rocky deserts in a craggy landscape with mountain ranges on the horizon. And yet, surprisingly, just outside the city, there are patches of green: plantations of young trees. This is where *qat*, Yemen's most popular drug, is grown. People chew *qat* leaves and store the masticated ball of greenery in their cheeks. And *qat* production devours a vast amount of Yemen's water.

Yemen's water problem is a typical overshoot problem. When diesel pumps became available, people simply drew more water and considerably expanded the land under cultivation. As long as any water made it to the top of the borehole, everything seemed just fine. No one was interested in what was going on subsurface in the water-bearing layers. Of course, nobody intended to destroy the natural systems. Today's scarcity is the unintended consequence of a classic if devastating side effect.

As soon as one's neighbor started to use a pump, the competition began. No one could buck it. In game theory, the situation such competition creates is called *the prisoner's dilemma*. A meaningful solution is only possible if all participants find the strength to let go of their competitiveness, coordinate their activities, and agree on a contract. In this case, it would require agreeing on an upper limit of consumption and on a system for allocating the available water. But once the impact of overshoot wreaks havoc, the willingness to cooperate can already be excessively eroded.

Yemen's water scarcity is a story without a silver lining. The metropolis of Sana'a and two million inhabitants cannot just be moved, and there may not be other places to move to. The pressure is mounting with population growth: the population of Sana'a has doubled every 10 years and that of Yemen every 20 years. All this is happening while Yemen's Footprint averages at only 0.7 global hectares *per capita*—and with the civil war has been shrinking. There is currently no obvious solution.

In this process, Yemen has quite closely followed the pattern of other low-income countries. Since 1961, their average Footprint of below one global hectare has remained practically unchanged. At the same time, their populations have rapidly increased and almost tripled: an almost vertical population growth. At 1.1 global hectares, the biocapacity of low-income countries lies significantly below the key figure of 1.63 global hectares that marks the world's average *per capita* supply in 2016. But Yemen has a biocapacity of a mere 0.4 global hectares *per capita*. Meanwhile, the resource gap between countries with low income and high income is widening further. The Footprint of countries that the World Bank counts as high-income and that now are home to a billion people has risen from 4.2 global hectares per person in 1961 to 5.1 global hectares per person in 2019. This happened on top of a 40% population growth in these countries.

Some of the lowest-income countries have transitioned from a biocapacity deficit into ecological bankruptcy. For example, Haiti's ecosystems are no longer capable of feeding the local population. At the same time, the country is financially and ecologically too poor to import the goods it lacks, especially in light of climbing food prices on the world market. Such situations cause hunger, social conflicts, and civil wars. Nearly a decade after the 2010 earthquake, the wounds remain visible. The country continues to depend on subsidized imports and privileged exports. The situation is similarly difficult in the Sahel (for example, in Darfur or Mali), made worse by ISIS conflicts. Regions such as these have been written off financially and play no part at the world's stock exchanges. For that same reason, they are also off the radar of Western media and carry no weight on the world stage. Their tragedies unfold hidden from view. The exception among these countries appears to be Pakistan and North Korea, which suffer from similarly difficult conditions but attract more Western attention because of their nuclear arms.

There is a saying, attributed to Alanis Obomsawin:[16] "When the last tree is cut, the last river poisoned, and the last fish dead, we will discover that we can't eat money." Many, though, think to themselves, "True. But the one with money will buy the last fish." At least that is how the current system is organized. But can we really rely on money? Does Earth contain enough resources to back up the paper value of money?

PART II

FOOTPRINT

Challenges Defining the 21st Century

CHAPTER 6

END OVERSHOOT!

Communication Is Key

From the Footprint perspective, the growing global overshoot is the 21st century's most fundamental challenge. Because Footprint accounts can adequately describe overshoot, such accounting serves both as an indicator and a management tool. The Footprint communicates easily and lets people participate in wrestling with what to do about overshoot. That's also why Footprint analysts do not arrive with a ready-made action plan. Rather, they work like coaches, enabling everybody to make their own decisions. With the Footprint tool, anybody can develop possible solutions toward creating, and subsequently monitoring, sustainable, robust economies. The goal of the Footprint is to end overshoot, and to do so in a humane way and not leave our fate up to the implacable pressures of the forces at play.

You may have seen the images of vast heaps of bison skulls slaughtered on the North American prairies in the 19th century. By the 1880s, they had been hunted down from herds of 50 million to just a couple of hundred. But there is an animal species in North American history that was hunted to an even larger extent and ultimately driven to extinction: the passenger pigeon.[1]

The passenger pigeon had a russet chest, white belly, and bluish-gray head and back. Its most important nesting grounds were in New England, New York, Ohio, and the southern parts of the Great Lakes. Flocks of these low-flying birds were so dense that people used to capture the birds often by simply standing on the crest of a hill and hitting them with clubs or by catching them in nets. A single shot could kill 30 to 40 birds. Some nesting colonies were 5 miles long and 12 miles wide, with single trees carrying up to dozens of nests to the point where many branches, and sometimes entire trees, collapsed under the birds' weight. Today's rough estimates suggest that by the mid-19th century there were about five billion passenger pigeons in North America. Afterward, their population steadily diminished.

And then, within half a century, the passenger pigeon was to be entirely wiped out. The necessary precondition for its extinction was a mass market for cheap meat, made possible by modern technology, the railway. From the early 1850s onward, the railway connected the birds' hunting grounds along the Great Lakes with New York and other big cities on the East Coast with their demand for meat. We have very detailed business records. On one single day (July 23, 1860), 235,200 birds were shipped from Grand Rapids, Michigan. Over the course of one year, 1874, Oceana County, also in Michigan, sent one million pigeons to the big East Coast markets. By the late 1880s, the once immense flocks of passenger pigeons had markedly thinned, and their decline would continue until the death of Martha, the very last of her kind, in confinement in 1914.

Why did people let the passenger pigeon go extinct? Why did people not stop the killing when it paid off less and less? Because the passenger pigeons were nobody's property, not even the government's. Because the cost of hunting was minimal—a horse, a net, a gun—whereas the prospects of profit were great. Many participated in the hunting and competed with each other. In this situation, it was advantageous for anyone to shoot more birds than others. And given the built-up demand, the faster the bird population was decimated, the more pressure there was to kill more, and to kill faster. For one person to hold back would have only meant that others got ahead.

Other ecosystems have not fared much better. Well into the 20th century, the vast oceans and their fish stocks were seen as simply inexhaustible. Today, huge factory ships fish on the high seas. Equipped with sonar, radar, and GPS, they get to the exact locations of the schools of fish and set gargantuan nets. Entire schools of herring, mackerel, or tuna are lifted from the water, processed, packed, and frozen on board. Such high-tech trawlers can annihilate fish stocks that survived traditional fishing methods for centuries.

In spite of ongoing technical upgrades to the fishing fleet, the overall amount of fish caught across the world has stagnated for years. Most fish stocks have reached or exceeded the limit of regeneration and are caught at younger and younger age. Again and again, fish stocks collapse, be it in the North Sea or on the Canadian Atlantic coast. In California's Monterey, it was the sardine fishery—made famous by John Steinbeck's novel *Cannery Row*—that collapsed overnight.

The same pattern of ruthless exploitation of nature can be seen in the histories of fur animal, seal, and whale hunting. Frequently, overexploitation led to the collapse of a stock and even the extinction of entire species.[2]

All these are examples of overshoot: the products of the planet's biocapacity are harvested faster than they regenerate. In this context, we often hear mention of "the tragedy of the commons," a phrase that is also the title of a 1968 article by ecologist Garrett Hardin.[3] His article is still the most cited paper published by the highly respected *Science* magazine, but it is also one of the most misunderstood ones. To a large degree, the misunderstanding is caused by Hardin's poorly chosen title. If Hardin had called his article "Tragedy of Open Resource Access" or perhaps "Our Common Tragedy," our understanding might be further ahead. Because, many of his critics, and Hardin himself, recognized that Commons are one possible solution to this tragedy.[4]

In his article, Hardin sketched out the following situation: several shepherds use the same large pastureland. Every one of them uses it as much as possible by taking as many animals as possible to graze there. For a while all is well. Tribal warfare, poaching, and diseases keep the numbers of sheep low. But eventually the limit of the pasture's biological capacity is reached. Each of the shepherds will ask themselves more or less the same question: What benefit will I get from adding one more sheep to my flock? The answer is clear: adding another one will increase their profits. But the end result is overgrazing, or overshoot. And that affects everyone. The shepherd has one more sheep, but everyone's now will be slightly skinnier—and the long-term prospects for this pasture to maintain sheep is also eroded through overuse.

Simply, if losses are socialized while profits are privatized, a shepherd acts rationally adding one more sheep to the pastureland, and another one, and still one more.... "Therein is the tragedy," said Hardin.

For grazing land to continue to flourish, the shepherds must agree on an upper limit for overall grazing, and (within those limits) who is allowed to take how many animals to pasture. An example could look like this: Every person in the village may take two sheep to the commonly owned pasture, whereas the third sheep is slaughtered by the community.

This is the kind of arrangement that prevailed with many commonly held resources throughout history. Examples include shared community resources, agricultural areas, fishing grounds, or water for irrigation in the Swiss Alps.[5] It is the same logic that international negotiations follow over the climate question. *Cap and trade* means essentially nothing but agreeing on a common and binding upper limit for greenhouse gas emissions and on a mode for allocating allowances to emit such gases as well as for setting fees (or penalties) for anyone exceeding their allowance.

Unfortunately, the discussion of how to manage common resources of large groups is often frustrating or moves at glacial pace, whether it is about the atmosphere, the world's oceans, or its forests. Hardin's article continues to shed light on why this is the case. The true origin, as Hardin described correctly, is free or uncontrolled access to resources. Everyone grabs what they can get, and no one assumes any responsibility.

At the core of commons management are two questions:
1. How much is there?
2. Who gets what?

For both of those questions, Footprint accounting can be useful. It helps to identify limits to regeneration for any ecosystem—from a field, pasture, forest, the biocapacity of a country, all the way up to the biosphere of the entire planet. Footprints and biocapacity assessments show the upper level of sustainable extraction above which overuse begins. The assessment then contrasts that overuse with actual use.

Any potential strategies—political, economic, or even military strategies—that may derive from its analyses are not the responsibility of Footprint accounting. As in the financial domain, a catalog of attempts at solutions is not part of accounting. But thanks to Footprint accounting, everyone can develop their own solutions and test their efficacy.

For the affluent, the overshoot phenomenon is often only an aesthetic experience. Through the window of the plane, we see how the cities below expand further and further: more houses, more highways and streets, more parking lots. Once it may have taken us half an hour to walk out of town and into the countryside to experience nature, whereas now it takes us twice as long. So we'd rather take the car to the forest where we will jog or ride our bike. Something similar occurs in tourism: once regions become impoverished, once their social tensions grow and they become unpredictable and dangerous, the affluent simply no longer travel there.

Generally, a loss of biocapacity is felt more directly by populations with low income. Take people in Kenya or parts of India: when their fields dry up, they have less to eat. As long as a drought happens only locally, the respective region loses purchasing power. But if production drops globally, the price of food goes up for everyone. People with enough purchasing power are affected only indirectly; they pay more for their loaf of bread but are hardly bothered. Others, however, can no longer afford sufficient food.

Overuse of biocapacity is typically recognized not as a systemic environmental problem but primarily as a sign of poor management, an unexpected drought, or some distribution problem; all of those can lead to tensions and conflicts. By not seeing the systemic connections, people may react fighting symptoms rather than rectifying causes.

But Footprint accounting shows that in almost all countries the demand for biocapacity has grown steadily. For sure during the past half century, according to the National Footprint and Biocapacity Accounts, which are based on detailed UN statistics stretching all the way back to 1961. With additional basic energy use and population statistics, the ever growing demand for biocapacity can be tracked back at least since the first industrial revolution. What becomes obvious is that the expanding human demand has been fueled by the discovery and growing exploitation of fossil energy.

All through the industrial age, when demand did not grow, political and business leaders as well as the public got quite concerned. Often economic stimulus packages were passed. But in our time, now, the supply, measured as *per capita* biocapacity, has moved in the opposite direction: it has sunk. Most people consider the current trend as "normal"; it is happening everywhere, and we, our parents, and even our grandparents have never known anything different. If today the demand for biocapacity rises everywhere while the *per capita* supply falls, it means we are all spending our capital.

Today the Earth's biosphere has a deficit of over seventeen planet-years. This means that humanity has used up over seventeen years of the planet's entire biological production.[6] Our ecological debts are accumulating. Science has yet to find the precise answer as to how long this global process of overspending can continue. But we know already the implications for portions of the biosphere. For instance, the accumulation of carbon waste in the atmosphere is already at a level that has committed humanity on a path of global warming beyond the 2°C established as the upper limit of atmospheric balance by the Paris Climate Agreement. Overshoot, already visible in various regions of the world, will definitely become an ever more decisive factor. Populations and countries with relatively little purchasing power and therefore little flexibility for adaptation will be its first victims.

The big question is this: Will this process simply continue? Or will we manage to reverse the trend? It is a gigantic challenge. The earlier we grapple with it, the better our chances. Much time has already been wasted. Once the ecological and social systems enter a state of chaos, turning the ship around will be exponentially tougher.

The core of the problem is obvious: Globally, humanity spends more nature than it earns. We are using up our capital. This cannot work in the long run. Yet assessing the ramifications of growing overshoot is difficult. One of the reasons is that people in high-income countries (such as Switzerland and Germany where we authors were born, or Canada and the US where Mathis now lives) have too many options in their day-to-day living to work around shortages. Many are doing well economically. But we do see clearly that not everyone on this planet can live this way.

The American researcher Jared Diamond has studied the conditions under which civilizations survive or fail.[7] His example of Easter Island stands out because it has obvious parallels with planet Earth: both are isolated, whether in the Pacific or in space. Both are on their own and are neither threatened nor helped from outside.

On Easter Sunday, 1722, the Dutch explorer Jacob Roggeveen came upon a remote island in the Pacific. Its beaches were lined with monumental statues with tall heads and strangely pointed noses. Most had been toppled and lay on the ground, shattered. All in all, it was a scene of devastation. The island's few inhabitants led marginal lives on a treeless pile of rocks at the end of the world.

Diamond suggests a scenario which may have unfolded. A thousand years before the Dutch arrived, Polynesians had come to the island in open canoes and settled there. At that time, dense forests still covered large parts of the island. The Polynesian settlers used the timber to build boats with which to hunt tuna and dolphin, and they used the bark to manufacture ropes. With these ropes, people managed to move heavy sculptures for religious worship—weighing up to 90 tons—on wooden sleighs. For special ceremonies, priests placed eyes made from white coral, with red coral for pupils, into the giant statutes' eye sockets. The statues' piercing, awe-inspiring gazes were aimed at the ocean.

And so the Easter Island residents continued to develop their prestige objects. They decimated their forest, both deliberately through logging and accidentally through the animals they had brought with them. Eventually the forest disappeared altogether, and so did the seabirds that had nested in its trees. Rainwater now ran unimpeded down the island's slopes and carried fertile soil into the sea. When there was no more timber to build canoes, there was also no more fishing. The Easter Island civilization was in trouble. Famines arrived and with them, some believe, cannibalism. The small remaining population was further reduced by diseases visiting Europeans introduced.

How was it possible for such a catastrophe to unfold on the island? The dozen or so clans on Easter Island may have started to compete to erect the most and biggest cult figures. People were blinded by their vanity and thirst for power. Clans blocked each other out of self-interest and were inescapably trapped in spiraling competition.

Their story may have unfolded differently. But in any case, it was a *tragedy of open resource access* in which the collective interest in survival was unable to win out over more private interests. Today, as the climate problem reminds us, overcoming such barriers is again of existential importance. A solution is only possible if we communicate with each other, come up with common rules, and ultimately agree on a binding contract that makes sure all abide by it.

Not too long ago, things were easy. People decided to build a road. Built it, done! Today we must take into consideration carbon dioxide emissions and a road's impact on things such as water balance or biodiversity in its vicinity. Footprint accounting summarizes all competing demands and tells us how specific human activities contribute to or subtract from overshoot. That way, Footprint accounts give us a statistical basis and starting point for the dialogue we need to have. Through dialogue, agreements, and ultimately action, we can peacefully resolve allocation issues, rather than let this conflict slide into violence and destruction.

The task of Global Footprint Network in cooperation with its partner organizations is to improve and standardize this tool and to make it accessible. The method must be protected from getting watered down or otherwise manipulated. To stay ahead of this game, we are taking these accounts to the next level of impact. Together with York University in Toronto, Global Footprint Network is creating a new Ecological Footprint initiative, with the goal of assembling a coalition of countries, supported by a rigorous global academic network, that owns and produces these National Footprint and Biocapacity Accounts independently and with highest reliability, so they can inform decision-making in an unbiased way.

But an understanding informed by the Ecological Footprint offers more: by looking at the world from the perspective of biocapacity, we gain a new, simpler, and more useful picture of our dilemma than just carbon or building merely on general concepts. Admittedly, humanity faces the enormous problems of a tragedy of open resource access: if I reduce my CO_2 emissions and as a result the climate benefits, I share this benefit with all of humanity. But the costs of my CO_2 emission reductions are mine alone. In contrast, if we take all the resource issues together—land, water, fossil energy sources, minerals—we find that many aspects, maybe even most of them, are not subject to this tragedy logic. To emit CO_2, I first have to buy fossil energy. If I consume less fossil energy, I am rewarded with lower costs. And if I reduce the energy dependence of my property, I increase its value. In spite of the financial crisis that temporarily reduced the fossil fuel price, these costs are on a significant upswing again, with increases possibly exceeding by far the carbon taxes that environmentalists have been hoping for (the only difference is that these increases are going into the wallets of the fossil fuel providers, not into the coffers of our own governments, which I am sure most people would see as preferable).

The argument that not every resource problem is subject to "tragedy" logic applies not only to energy. If a country eats more food than its farms can produce, this extra food has to come from somewhere. Such a food risk is borne by the country alone. Again, the biocapacity perspective—with the portfolio of all resources and ecological services that require area—is useful in demonstrating that it is in any country's interest to moderate its hunger for resources.

Also, the fact that much of our resource consumption is dictated by the infrastructure of the place we live in reinforces how important it is for us to think about our own interests. If our country, our city, or our company has an infrastructure that is excessively hungry for resources, our own economic risks go up. The common climate conversation does not yet sufficiently recognize this relationship. Could it be possible that if we do prepare for a possibly turbulent future of growing ecological constraints, it is we ourselves who will be prepared? The next chapter delves into exactly this topic.

CHAPTER 7

WINNERS AND LOSERS

Strategies for Countries to Consider

In a world increasingly shaped by resource constraints, the conditions of global competition are changing rapidly. Under the old rules, countries and regions tried to attract as much financial capital as possible even at the risk of harming their social and ecological capital. Everyone placed their bets on economic growth. But in a future shaped by overshoot, prudent use and careful conservation of one's ecological capital will be far more important than spending it. Biocapacity deficits are becoming more and more risky: they may become expensive, or if prices do not react, disruptive. Natural resources are turning into key factors for economic competitiveness. With rising global pressures, countries and regions need to position themselves wisely. Their most fundamental interests are at stake.

Ecological Footprint accounting helps to gauge opportunities and risks and to develop individual strategies for the future. Efficient infrastructures for mobility, housing, and energy supply are crucial. Perhaps we are not mired as deeply in a tragedy of open resource access as many believe. Perhaps the problem becomes more solvable if we focus more on our own situation. Let us start with Switzerland.

People from many cultures and continents love *Heidi*. It was in 1880 that the Swiss author Johanna Spyri published that children's book. Since then, little Heidi—with her dream of unspoiled nature—has conquered the hearts of millions of people. In the book, Heidi is

In a world that is overstretched with human demands,
a country's access to sufficient biocapacity is becoming ever
more significant a determinant of a resilient economic future.
This underlines the significance of every country's biocapacity
within the country's legal domain, i.e., its "farm size."

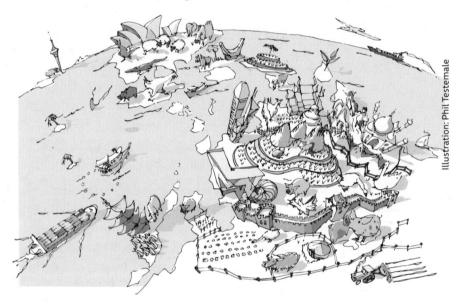

Illustration: Phil Testemale

an orphan. After her parents' death, she is sent to live high up in the Swiss Alps with her old, embittered grandfather. Together with her friend Peter, she tends the goats. For both children, nothing more beautiful exists than the untouched Alpine world. One day, though, Heidi is sent to live with a family in the big city of Frankfurt in distant Germany. When, at the end of the novel, Heidi returns to her grandfather, she has experienced the noise and the hectic way of life in the big city with its technological civilization and now sees her home in the Alps with new eyes—and loves it much more deeply.

As conservative as it may be, Switzerland has changed in the years since 1880. But Alpine huts and clear mountain streams, forests, and pastures with glorious wildflowers still exist. Despite their industrialization and modern civilization, the Swiss appear to have successfully hung on to big parts of their magnificent nature, almost as it was in Heidi's day.

By 1961, the first year with complete Ecological Footprint accounts, Switzerland's Footprint exceeded its biocapacity by 220 percent—or Switzerland used 3.2 Switzerlands. It exceeded its biocapacity for quite a while before (but we have not sufficient data to calculate for how long before). Right up to the first oil crisis, the country quite rapidly increased its hunger for resources, especially energy, as is typical of industrial societies. By 2016, Switzerland used 4.5 Switzerlands. There is a silver lining: From 2006 to 2016, its per person Footprint reduced by one hectare from 5.67 to 4.64 global hectares per resident, mainly due to decarbonization and a reduction in the consumption of forest products. Wood is the only biological resource the country, in net terms, has enough of; after all, 31% of the country is covered with forest. Hydro power supplies 55% of Switzerland's demand for electricity. For their food supply, the Swiss depend on massive imports of biocapacity. Their food Footprint exceeds their domestic agricultural biocapacity 3.5 fold.[1]

Most of its cattle no longer graze on Alpine pasture as in Heidi's day. To reduce nitrate pressure from high cattle concentration, farmers in Switzerland are now encouraged to less intensively graze and increase plant biodiversity on Swiss meadows. But in response, Swiss residents have not reduced their demand for meat or dairy. Therefore, more of the beef consumed by the Swiss is now raised in South America. Also, given biocapacity constraints in Switzerland, its cows now get additional feed from abroad, including South American soy meal. Meanwhile, the tropical rain forests in South America keep shrinking. Unfortunately, cattle cannot easily digest soy meal. They tend to get bloated and release methane, a powerful greenhouse gas.

As a result, because of the reduction of alpine grazing, the Swiss Alps now carry less of an ecological burden, and their low-nutrient meadows with their wildflowers shine in renewed glory. It looks perfect in Heidi land. Hardly a Swiss resident, let alone any tourist, realizes that the Swiss landscape and lifestyle are massively subsidized by imported biocapacity and pressure on biodiversity elsewhere.

Not many countries can afford such luxury. But thanks to improved ecological efficiency, Switzerland's total Footprint is now rising more slowly. Still, it far exceeds the country's capacities, and its per person Footprint is markedly above the global average. The trajectory into the future is not clear as its population keeps increasing, while its *per capita* biocapacity continues to slowly drop.

The Swiss government has recalculated its Ecological Footprint results a number of times. Its latest review in 2018 again confirmed Global Footprint Network's results. The government study reproduced the National Footprint and Biocapacity Accounts' time trend within a narrow band of less than 3%.[2]

Global Footprint Network is now covering over 200 countries, and a number of countries have gone through verification research confirming the Network's predicted trends, even when using their own rather than the UN data sets.[3]

The pictures emerging from country assessments are quite varied. Take Morocco, for example: its *per capita* Footprint is significantly lower than the global average, and began to rise from 1.3 global hectares per resident in 2000 to 1.8 in 2009—and has stayed flat since then. But Morocco's biocapacity is also significantly lower than the global average and is subject to strong fluctuations due to its dry climate. Drier years translate immediately into lower biocapacity. Or take Tanzania: its *per capita* Footprint has shrunk. But because of strong population growth, its *per capita* biocapacity is now below the global average. You can take a look at your favorite countries by consulting Global Footprint Network's data open platform.[4]

Because of its size and dynamic development, China is of particular interest. Landing at Beijing airport today and entering its new Terminal 3 would take anyone's breath away: the world's largest hall, designed by British star architect Lord Norman Foster. But equally breathtaking is the air outside Beijing's palatial airport. Despite every effort to improve the capital's air quality for the 2008 Olympics, this metropolis with its millions of residents still suffers often from a haze of industrial and traffic fumes, in spite of impressive transformational progress: such as replacing all two-stroke engines with electric motors, and cleaning up and closing down the most outdated industrial polluters. Residents of other Chinese conurbations in more rural areas are burdened with even more pollution. According to the Global Burden of Disease 2015 study, air pollution is thought to have contributed to the premature death of 1.1 million Chinese. (Nevertheless, over the past 20 years, life expectancy in China has grown by 5 years for males and 7.5 for females.)[5]

Even though its growth has slowed down, China's ecological fate is closely tied to the planet's because its population is so large and because the country is pursuing a path of steep economic development. While the Chinese residents had on average a Footprint per person of below 1 global hectare in the early 1960s, that Footprint had risen to 3.6 global hectares by 2016. This means that China's Footprint per person is now about one global hectare higher than humanity's average Footprint, far exceeding the biocapacity the planet can provide to each person. China's biocapacity adds up to about 1 global hectare per person, significantly lower than the 1.63 global hectares per person worldwide in 2016 (and this includes the biocapacity required by wild animal and plant species). As a result, China has had a biocapacity deficit since the mid-1970s and now consumes ecological resources of more than "three and a half Chinas."[6]

The country covers parts of its deficit by importing natural resources at the rate of 290 million global hectares (in 2016). That number roughly corresponds to the entire biocapacity of Germany. A large part of its deficit is the result of China's demands on the atmosphere, a resource shared by the planet. The natural resources China imports come predominantly from countries far away, such as Canada, Indonesia, Mozambique, and Brazil. Some of China's resource demand gets passed on through the export of its products, as China has become the world's workbench. But surprisingly, according to the National Footprint and Biocapacity Accounts, China is now importing more biocapacity than it exports: China's production Ecological Footprint is 0.2 global hectares smaller than its consumption Footprint, suggesting that it imports more biocapacity than it exports. Put differently, the biocapacity embodied in everything imported surpasses the biocapacity embodied in everything exported by China.

Will China be able to change course in its race to catch up economically? Will it manage to not copy the highly resource-intensive patterns of production and consumption that characterized industrialization in Europe and the US? Or will it be able to transform to a sustainable economic development and pursue a path of eco-efficiency and social stability? China seems to realize that such a path would be in its own best interest. At least China's official rhetoric does face up to this challenge. There is talk of a "circular economy" and an "ecological civilization." Resources and ecology are concepts prominently mentioned in China's five-year plan. Also when looking at the latest resource trends, China's Footprint seems to be flattening, after two decades of rapid expansion. Obviously, the verdict is still out as to whether China can go far enough in its change of course. Global Footprint Network collaborated with one Chinese provincial government (Guizhou) to shed light on this challenge.[7]

Footprint accounting helps to find solutions. Whether China or Switzerland, Morocco or Tanzania, every country has its own risk profile and unique development trends. The following questions help to assess a country's situation:

1. What are the Footprint and biocapacity trends of this country, among its main trade partners, and in the world at large?
2. Does this country run a biocapacity deficit or does it have an ecological reserve? If it runs a deficit, does its financial purchase power grow fast enough relative to other countries' for it to be able to buy any missing biocapacity?
3. What natural assets does the country have? Is its biocapacity— or its biocapacity reserve—growing or shrinking? What are the reasons?
4. Does demand grow faster than technological efficiency? Or what else may drive the country's Footprint trends?
5. What are the country's ecological risks? How fast are the trends moving? What might be the effects on its competitiveness, today

and tomorrow? What is the respective situation in countries it has partnerships with, especially those it imports from and those it exports to? In other words, it is not only the country's ecological performance that matters, but also that of its trading partners, particularly for the country that significantly depends on biocapacity from abroad.

6. What is the country's ideal situation, i.e., its optimal consumption of natural resource? More precisely, what is the country's optimal balance of resource consumption (and CO_2 emissions) on the one hand and biocapacity on the other hand, given that too much consumption becomes a risk while too little may make living well more difficult?

7. Have all opportunities been taken advantage of to live better lives with fewer resources?

8. Are the country's infrastructure investments designed to make the country less resource dependent? Or will they make the country more vulnerable to global resource risks?

These questions can help to develop national solutions. Beyond that, Footprint accounting can monitor trends to test whether actual efforts produce the intended directions.

Take transport systems, for example. The concern is widespread: If one day the Chinese drive as many cars as Europeans or Americans, what then?[8] Let's do the math. There are 1,400 million Chinese, 200 million of whom have moved to the cities over the past decade. On average, they consume about 70 liters (or 18.5 gallons) of fuel per person per year. The bottom line is that they produce markedly less greenhouse gas than the residents of Atlanta, Georgia, or the residents of Charlotte and Raleigh, North Carolina. In those American cities, the annual per capita consumption of fuel amounts to 4,000 liters (or slightly over 1,000 gallons).

Really? That much? When it comes to fuel consumption, these three cities are the frontrunners in the United States. New Yorkers, for example, fill their tanks with less than half that amount. In European cities, the average consumption per person and per year is roughly

450 liters (or 120 gallons). Atlanta deviates far from the more common numbers because the city was designed around cars. Its population density of six people per hectare is miniscule. Walking won't get you anywhere in Atlanta, but the city's road network is gigantic.

In modern Chinese cities, 150–200 people live on one hectare. For them, area limits road construction. Chinese cities are not destined to become like Atlanta. That does not mean, though, that the wave of motorization in China and other parts of Asia will not accelerate China's development.

What becomes evident is that a city's layout, especially that of its transport systems, enshrines its resource consumption for decades if not centuries. People who live in Atlanta, go to work there, take their kids to school, or do their shopping will need their 1,000 gallons of fuel per year, whether they like it or not. And it won't be possible any time soon to completely rebuild the city, as had been necessary after its total destruction in the American Civil War. And even if it were possible, it would require enormous resources, as became obvious in New Orleans after Hurricane Katrina in 2005.

But what does work, and works very well, is building within an existing city a state-of-the-art local transport system that moves people more conveniently than cars do. There are many examples on all continents. A prominent case is Paris, which has high population density and in recent years has pushed back against vehicle traffic to make more room for pedestrians and cyclists and to create a new express bus system. This has definitely been wise planning by Paris's municipal government. Cities and conurbations that today decide to build strong and energy-efficient rail or bus systems and reduce vehicle traffic will be tomorrow's winners. Also, emerging electric scooters and bicycle share programs carry a lot of promise.

Marriages may be "forever"—yet one can leave a marriage or get a divorce. It is not that simple with infrastructure. Infrastructure will stay in place for years or decades. Road systems leave traces for centuries—or even millennia, as we can see from cities built by the Romans. We cannot rebuild our infrastructure at the last minute. Well-chosen infrastructure such as livable, efficient cities is a gift,

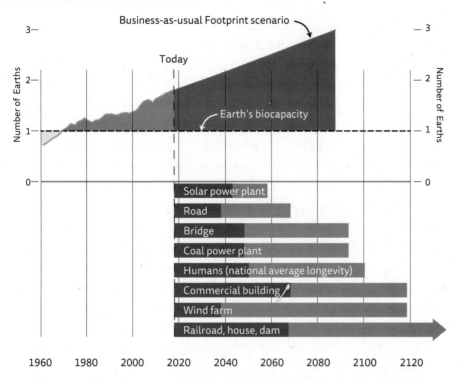

Figure 7.1. The various ranges of longevity of people and new infrastructure investments, within the context of humanity's future Footprint. As humanity's annual biocapacity deficits accumulate, infrastructure put in place today will have to operate in a world of an ever more massive biocapacity debt. Which infrastructure will become increasingly useful in such a world, and which will become a burden? Credit: National Footprint and Biocapacity Accounts, 2019 edition.

a valuable legacy. In contrast, obsolete infrastructure such as coal power plants, inefficient housing, or dispersed suburbs become expensive or simply a nuisance.

This is why it is so important that states or cities make their infrastructure decisions and investments in the light of future conditions. The key to sustainable development is building only infrastructure that is one-planet compatible, and retrofit anything that is not. Global Footprint Network simply calls this: *Slow things first!*

As global uncertainties mount, it is increasingly important for every country and every city to prepare. If, for example, neither China nor India, nor the United States or any other of the economic heavy-weights budge on the climate issue, things will get tight, especially for smaller countries. Biocapacity deficits will become increasingly more dangerous. Ensuring that one's natural capital retains its ability to regenerate or, better still, to grow will become ever more important; keeping a lid on demand will also be critical. And old questions will gain new urgency: How can we maintain our quality of life with our existing biocapacity? How can we position ourselves so as to protect ourselves from being sucked into a downward spiral?

Against the backdrop of increasing water scarcity and desertification, worsening erosion and soil loss, diminishing per hectare agricultural yields, overgrazing and destruction of tropical rain forests, species extinction, overfishing, and climate change, the rules of the game in the global competition change. Energy adds pressure to the game, as climate change puts even more limits on the use of fossil fuels, far more than the fossil reserves underground limit the use. Waiting for geologically imposed peak oil (which would eventually be inevitable) will be far too slow. We need socially imposed peak oil and rapid phaseout to preserve a stable climate, and with it reliable biocapacity. From this perspective, what we are facing has been more aptly described as *peak everything*.[9]

If we consider the carbon dioxide problem in isolation, the call for action is ambiguous. After all, I know that my personal reduction of carbon dioxide emissions will do little to help the planet's climate since I am only one of over 7.7 billion. In this regard, the game we play of who has the right to emit how much is like the previously mentioned prisoner's dilemma or the so-called tragedy of the commons: As long as everyone is seeking their own advantage, everyone will lose. We are all residents on a planet whose atmosphere has long stopped meeting our demands. This leads to the self-defeating strategy of, "Only when all participants have joined hands can I risk taking my own first step toward reducing my carbon dioxide emissions." To approach the climate problem as an isolated concern becomes paralyzing and unproductive. Everyone waits for everyone else.

If, on the other hand, we approach the challenge from a peak-everything perspective, then each country is really in its own driver's seat. After all, with all the available climate science, it becomes obvious that the only viable path forward is an economy that lives off our planet's regeneration, rather than its liquidation. The choice is merely how fast we get out of liquidating fossil fuels and, with it, Earth's climate stability. We can go faster and comply with the Paris Climate goal, thereby rescuing much of our precious biocapacity (this means ceasing fossil fuel use well before 2050). Or we can go slower, still having to eventually move out of fossil fuel use, but then being left with less, and less reliable, biocapacity. Either way, it becomes clear that climate action is a question of resource security.

In other words, if your city or your country is not already preparing for the inevitable future, your city or country will not be ready for that future. And that future will not be pretty. It is that simple. There is no benefit in painting yourself into a corner by continuing expanding resource-dependent infrastructure and resource-gobbling economic sectors. We are not stuck in a so-called tragedy of the commons—rather we are stuck in believing that we are trapped in a tragedy of the commons, stupidly waiting for others to act first.

Ecological conditions are crucially important but may vary dramatically from one region or nation-state to the next. For example, Sweden with its Footprint of 6.3 global hectares per person and its biocapacity of almost 10 global hectares is still an ecological creditor even though the Swede consumes on average almost four times more than is available per person on the planet. Bangladesh, on the other hand, has a Footprint of only 0.9 global hectares but is an ecological debtor because the biocapacity of the country is only 0.4 global hectares per person.

What are you waiting for? Move early and quickly if you want to have good prospects. Ultimately, wise management of each country or city's resource security not only benefits each member of the human family but also enables the countries and cities which act to thrive.

Such management though presupposes reliable ecological accounting. We will take a decisive step forward when government ministers cease viewing ecological questions primarily as cost factors and instead embrace them as offering crucial competitive advantages. We will know that they have realized what is at stake when growing biocapacity deficits make those very ministers break into a cold sweat the same way rising unemployment figures do. At that point, they will also recognize investments in sustainability as opportunities and indeed as indispensable for their country's future.

Throughout the 20th century, and still today, society's fixation has been: generate more income at all costs! That's the essence of the growth fixation. Countries strive to attract as much financial capital as possible, even if it means harming its social or ecological capital. This focus no longer serves the 21st century. It would be far wiser to focus on maximizing your wealth. Wealth is the sum of all your assets that are able to provide you or help you generate future benefits. With biological capital being the foundational ingredient of all activities and being depleted fastest of any asset class, this means you need to manage your biological capital, or more precisely your resource security, with extra care. Let resource security give you the competitive edge.

Put another way, instead of consuming as much biocapacity as possible (which is not the goal but an unintended consequence of maximizing income and consumption), countries need to refocus on protecting their assets, particularly the biological ones. In a resource-constrained future, countries with the strongest biological capital, will have the upper hand.

This stands in stark contrast to the conventional concept of development, including within the UN, that continues to label countries as "developing" or "developed/industrialized." This label is neither descriptive nor explanatory. It is merely a thoughtless and destructive endorsement of GDP fetish. In reality, there are not two types of countries, but over 200 different countries, all faced with the same laws of nature, yet each with unique features. It is limiting and patronizing to act as if there were only two distinct sets of situations.

What is particularly absurd is the implicit assumption that the only path forward is striving for high incomes, while in reality those high incomes as encountered in much of the US, Japan, Europe, or the gentrified parts of Shanghai or São Paulo lead to massive Foot-prints that if extended to everyone would by far exceed the capacity of three Earths. In other words, conventional expansionism promotes a development perspective that is not replicable and therefore de-structive. This is not a new insight. When a smug British journalist asked Mahatma Gandhi, just as India gained independence, whether now, unchained, India will catch up with British levels of consump-tion, Gandhi apparently responded, "If it took England the exploita-tion of half the globe to be what it is today; how many globes will it take India?"[10]

Many problems, including income gaps, cemented our belief in GDP growth as the panacea for all ills. GDP expansion allows the privileged to circumnavigate the delicate issues of redistribution. We can appease the underserved by promising more for all in the future. However, given the large level of planetary overshoot, it is now far more important to ask whether a country has slipped into a biocapac-ity deficit or still has extra capacity. Secure access to biological capital is increasingly becoming the ticket to a prosperous future.

If your analysis requires that countries be organized by income groups, then keep the differentiation descriptive and call it *organized by income levels*. Don't interpret by calling low income "developing," "South," or "Third World." This is not out of political correctness but because these labels obfuscate and confuse, create a false dichotomy, foster lazy thinking, and encourage misguided policies.

Nearly as misinformed is it to use the adjective "poor" for low-income and "rich" for high-income (as most UN institutions and the World Bank do). Rich and poor refer to wealth, not income. Wealth is a stock, income a flow. It is like confusing speed and distance. Or population size and population growth. A further question is poor in what? Poor in resources? culture? biodiversity? gold? ideas? Let's use the power of science and be descriptive, rather than blindly perpetu-ating prejudice.

A descriptive distinction that might become increasingly helpful is to differentiate countries with biological capacity reserves from those who run deficits. Just take the world's biocapacity creditor giants today. The largest one is Brazil, far ahead of Canada, Russia, Australia, Congo, Bolivia, Argentina, and Colombia. Given growing overshoot, all of these countries have extraordinary opportunities if they manage them well.

Without paying sufficient attention to the resource realities, they may not organize themselves well, and fail to seize their advantage. Their economic strategists may not recognize the ecological power they possess. Also, having a lot of biocapacity or running a biocapacity reserve does not mean that they use their ecosystems carefully and wisely (much indicates they don't). It just means that they are luckily endowed with a lot of biocapacity. If we Ecological Footprint analysts were they, we would ask ourselves in which markets we would gain strategic advantages? If they paid attention, they would see that already today the geopolitical lines are shifting.

The world's biocapacity debtors, on the other hand, include (among others) the US, China, India, Germany, Switzerland, France, or the United Arab Emirates. These countries are not only vulnerable but in the future will very likely have to pay ever higher prices for the fruits of other countries' biocapacity. Or they may not be able to get enough and face disruption.

Financial markets are also rethinking priorities. Analysts are beginning to realize that they have underestimated ecological risks. Some act on their moral convictions so as not have their investments contribute to overshoot. Others are more pragmatic. If Mexico has less biocapacity than Brazil, where are the opportunities, and where are the risks? What's the better place for an investor to seek opportunity? Global Footprint Network worked with the UNEP Finance Initiative on those very questions.[11]

The vast majority of today's high-income countries have had a clear historical advantage: they were able to attract financial capital and construct built capital before natural capital became constrained. At the same time, they managed to use technological advantages to expand their economies. On the basis of this position, they are able today to buy other countries' natural capital. Today, countries that are building out their industrial base on the same economic model (such as China and India) are massively expanding their infrastructure—roads, railroad lines, power plants, factories—all of which demand huge expenditures of material and energy; these projects cost vastly more now than they did 100 or 150 years ago.

During the second half of the 20th century, global production of goods grew by a factor of seven.[12] In the same period, global trade grew, in value, even more: possibly by a factor of thirty.[13] Material flows continue to expand. Global trade, fueled by modern transport technologies and the belief in free trade, is the main engine driving the depletion of natural capital. This development certainly has had its benefits—many live longer and more comfortably—but simultaneously we with high incomes have become dependent on flows of resources that cannot be sustained in the long run.

Financial capital can be moved around the world with the click of a mouse, while freighters and cargo planes do the rest. Natural capital moves in entirely different periods of time. Forests, for example, need decades to regenerate; oceans, centuries. Were it not for nature's strength and resilience, we could not overexploit Earth today at the level we do. But there will be consequences. Nature has a good memory, and damages accumulate. One day, humanity will get the bill.

In this situation, countries and regions can avail themselves of different strategies for increasing their resource security.

CHAPTER 8

FOOTPRINT SCENARIOS

Ways out of Global Overshoot

Ending global overshoot by design is a gigantic task with no historic precedent. It is made even more challenging as the human population is still growing, and as many people living in lower-income areas still heed legitimate material aspirations. Footprint accounting gives us context and metrics to evaluate pathways. Cities, regions, and countries need to identify their optimal level of resource consumption. How much biocapacity is available locally? How much, globally? How big is their purchasing power compared to that of others? Anyone who consumes too much puts their economic future at risk. If they consume too little, life may get less comfortable. If we all learn to steer our resource consumption more deliberately, and are committed to it because we recognize the inexorable link to our self-interest, we will have a far better chance of creating a world where all can thrive within the means of our one planet.

We cannot know the future. What time frames are plausible to pull humanity back into a stable situation? What paths present themselves? Which ones are realistic? Scenarios depict paths and contextualize actions and interactions; in a way, they shine a light onto possible futures. In doing so, they show us options and their potential consequences. They are especially worth exploring if the path ahead is unclear yet we want to act rationally and—against the odds—with foresight. Let's call these scenarios ways to stock up on ideas.

Scenarios function a bit like improvisational theater. The opening position remains the same, as does the scenery. But there is no text, no screen play. All there is is a framework: a setting, an agreed-upon duration, and a selection of characters. These start to create a story. How it will end on any particular evening remains to be seen.

Footprint accounting shows the opening position: the state the biotope is in and how it is being used. The ensemble has been selected, the number of actors determined. The stage is set, the house lights are going down, and the curtain rises. The play can begin.

But stop! Before it starts, let us prepare ourselves for what we are about to see. What does a Footprint scenario tell us about real life? What can it depict, and what not?

Footprint accounts provide us with a pair of glasses that filters out certain aspects as we look at the world. Diseases, for example, have no Footprint. It is possible to calculate the Footprint of, say, a hospital, but not of a disease as such. Neither do Footprint numbers tell us whether we live in a beautiful city or an ugly one. What's more, the lenses are ground to focus our eyes on the services provided by self-regenerating resources: the basic things life requires. The Footprint does not aid in our desire to find happiness in our personal lives. Maybe that would be asking too much anyway. But we most certainly—collectively but also individually—need certain things to sustain us in the first place and to give us a chance at fulfilling lives: food to eat, clothes to wear, a roof over our head. Footprint accounting has a good eye for those things, is scientifically verifiable, and at the same time very simple and graphic to allow for easy comparisons. It focuses on the material conditions our lives depend on. Footprint accounting compares human consumption of nature with the planet's biocapacity.

One more thing, and maybe the most important one: The number one rule for the scenarios is "One Planet Living." As far as we know, there is only this one planet in human reach able to produce chocolate; let's make the most of it! Life is meant to be wonderful. And all human beings ought to have a chance to thrive and live with dignity on this Earth.

The following two scenarios are designed for the whole planet. They are based on studies by the International Panel on Climate Change (IPCC) and the Food and Agriculture Organization of the United Nations (FAO). On the supply side, they ask: What will agriculture be able to produce? On the demand side, everything is more dramatic. How much food will humanity need over the next generations? How much greenhouse gas will humanity emit? The time horizon of many scenarios ranges as far as to the end of this century because many factors, such as population size, do not change rapidly.

The year 2004 was the first time we sketched out global Footprint and biocapacity scenarios. Some were rough sketches, others more detailed. One set of scenarios we developed in cooperation with the World Business Council for Sustainable Development (WBCSD) and 30 of its member companies. Those scenarios became the context for their Vision 2050. For the Living Planet Report 2012, which we prepared for Rio+20 (the UN Conference on Sustainable Development in Rio 2012), we simplified scenarios we had developed in 2008. All of our scenarios identify possible paths we could take, but the actual path humanity has been following so far has matched in every single scenario that most resource-intensive choice: "business as usual."

We prepared the two scenarios published here for Earth Overshoot Day 2015. They compare a "business as usual" path with a path that, according to the IPCC, is reasonably probable to limit global warming to 2°C.

Scenario One: Conservative Estimates
If Current Trends Continue

For this scenario, we adopted the conservative projections of the United Nations: slow population growth (as before), moderate energy consumption (as predicted by the International Energy Agency), and an increase in agricultural productivity (at the same rate as during the past four decades)—all in all, a simple continuation of existing trends. The result is startling. The gap between Footprint and biocapacity widens beyond all bounds and within decades assumes dimensions that are physically almost impossible.

Also note that, since we generated this scenario for the first time, population projections by the UN for 2100 have been upward corrected, particularly for Africa. The fossil fuel share in the overall commercial energy mix has not dropped. Carbon emissions in 2018 may have increased a staggering 2.7%, after also increasing 1.6% in 2017.[1]

Already by 2030, in this scenario, global demand for cropland and land that can absorb CO_2 will have climbed well above today's data, and our Footprint will have expanded to two planets with a trend to grow even further.

This will push our ecological debt from today's 17 planet-years to 28. All of that in only 11 years! If we extrapolate the curve further, assuming constant linear demand expansion as we have seen it since the 1970s, humanity would have reached 50 planet-years worth of ecological debt by 2050. Consumption will rise to the point of requiring 3 planets by 2060. At that time, our ecological debt will amount to more than 100 planet-years. One planet-year is the equivalent of what our biosphere produces in one year. So 17 planet-years could be compensated for if we were to consume nothing at all of nature for 17 straight years and if we also (optimistically) assumed that the whole overexploitation were reversible. But what really does 50 planet-years mean?

Let's paint a picture: It takes half a century for a healthy forest to grow. Annual growth rates for a young forest's accumulated biomass are about 2%. Hence we can take out 2% of a forest's biomass every year without damaging its ecosystem. But of course we could also log the whole forest in one go. The forest would be gone, and if the soil were not destroyed and new trees planted, it would again take fifty years before we had another mature forest. Forests are the ecosystems with the biggest accumulation of living biomass. Grazing land and cropland have only the soil as a stock (which can get quite substantial). Oceans contain only as much biomass as can be regenerated in eleven days. Because of the lower stocks in pastureland, fields, and the sea, these ecosystems are less capable of buffering large levels of ecological debt.

The atmosphere, too, has absorbed considerable quantities of our "CO_2 debt." This is both a gift and a trap because its acting as a buffer slows humanity's corrective reactions. Without immediate feedback, we merrily continue to pile up ecological debt. How much debt can the atmosphere swallow? According to the National Footprint and Biocapacity Accounts, 35 gigatons of CO_2 emissions per year correspond to the planet's entire biocapacity—or one planet-year worth of production. Human-generated CO_2 emissions have risen from 24.5 gigatons in 2000 to 36.2 gigatons in 2017.[2] In other words, those carbon emissions alone now exceed the entire biocapacity of the planet. The result of these annual emissions is a two to three ppm increase per year in the CO_2 concentration in the atmosphere.[3] Pre-industrially, the atmosphere contained 278 ppm CO_2 equivalent, which has risen in 2018 to 496 ppm.[4]

Let's contrast this with findings of the International Panel on Climate Change (IPCC), which is the UN's effort to summarize the most commonly agreed, peer-reviewed, and published science findings in the climate research literature. This institution brings together hundreds of climate scientists to produce large-scale reviews of the literature and summarize them in the so-called Assessment Reports. The last one, number five, came out in 2014. The sixth one is scheduled to be released in 2021. The fourth report included the recognition that a 450 ppm CO_2 equivalent concentration would give humanity a 66% chance to never increase the global average temperature 2°C above the preindustrial average temperature.[5] In other words, 450 ppm CO_2 equivalent is weaker than what the Paris Agreement stipulates. In addition, some renowned climate scientists argue for 350 ppm CO_2—hence also the climate advocacy organization with that very name: 350.org.[6]

Simply put, there is no budget left for carbon emissions. To comply with the Paris Agreement's climate goal, humanity will need to move out of using fossil fuel very fast—well before 2050—while also providing additional carbon sequestration. In essence, there is no real room to accumulate more biocapacity debt from the carbon Footprint without lasting damage to biocapacity.

Here is another way to explain the implications of planet-years worth of ecological debt. Consider this: If the whole planet were covered with mature forest, the world could amass a debt of 50 planet-years and might yet recover. However, then the planet would be clear-cut, and we would have to wait another 50 years for the next harvest. This means we would not be able to harvest anything in the meantime if we want to repay the debt.

Given that we are already at seventeen planet-years of debt, that the atmospheric buffer is already full, and that the entire planet is not grown over by forests ready to clear-cut, expanding this debt seems beyond risky.

This scenario of "Conservative Estimates If Current Trends Continue," which the United Nations implicitly appears to have in mind, is thus highly unrealistic. In addition, we need to consider that not everything in ecosystems proceeds in a linear fashion and that damages can lead to tipping points as planetary boundary research has emphasized.[7] For example, Earth may come to a tipping point if the Amazon River dries up or if too much global warming leads to a thaw in permafrost regions and a sudden release of enormous quantities of methane. Both events would speed up climate change and put even more pressure on the planet's biocapacity. In either of these cases, humanity would experience a considerably faster loss of biocapacity—and well before our natural capital was entirely consumed. The massive bleaching of reefs, caused by higher carbon dioxide concentrations in oceans (leading to acidification), ocean warming and eutrophication, are already well under way, also reducing if not eliminating the productivity of these amazingly diverse ecosystems.[8]

All the human demands on ecosystems do not exist in isolation. Gaining more cropland often occurs at the expense of forest. A dramatic example is the development of palm oil plantations, with particular prominence in Indonesia, but equally happening across the tropical belt, to the detriment of the country's rain forests.

As forests are converted, not only is biodiversity under assault but fewer trees can be used for timber, paper production, or carbon dioxide absorption. When fishing grounds collapse, pressure on

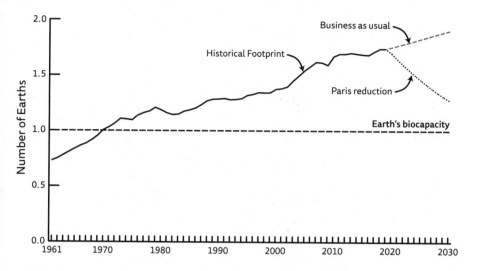

Figure 8.1. Two possible scenarios for humanity's Footprint: Rapid decarbonization (as prescribed by the Paris Agreement) or "business as usual," in number of Earths.

cropland tends to increase to supply humans and their pets with land-based protein. In short, an abstract calculation of ecological debt in planet-years cannot help but underestimate the dangers of growing overshoot. Such calculation serves only as a general indicator for risk. Most likely an underestimate.

The outcome of this scenario is unequivocal: An additional continual rise in our demand for biocapacity is beyond anything the planet can sustain. It is a destructive path. Besides the social, cultural, and likely also military tensions that will be caused by "business as usual," the biggest and most lasting danger will arise from the erosion of biocapacity. Essentially, the scenario of "Conservative Estimates If Current Trends Continue" shows us life on a planet that is rapidly becoming desolate.

In summary, the basic idea of this scenario is simple: As we compare the demand side implicit in official projections with what the planet can renew, Footprint calculations make obvious the material implications. What may come as a surprise: Even conservative scenarios do not come close to meeting the minimal conditions of sustainable economies in which the production of goods and services is matched by what the resources can provide. Our collective confidence in "business as usual" flies in the face of physical reality. To be blunt: humanity has failed to reckon with reality.

The second scenario gets serious about the demand to end overshoot. With the help of the Ecological Footprint, we will go on a journey of discovery through unknown territory to test its viability for finding a sustainable path through the 21st century.

Scenario Two: Reduction as Targeted by the Paris Agreement

This target is to have gradually phased out our overexploitation of the planet by the end of the century. Abrupt development should be avoided as much as possible. In concrete terms, this means that our ecological debt will still grow for a while in order to then—soon—shrink. A core issue is that of energy generation and carbon dioxide emissions. Our global carbon Footprint has tripled since 1961. And currently it forms by far the biggest single component of humanity's Footprint. It is therefore the component that this scenario focuses on.

The Climate Agreement negotiated at the 21st UN World Climate Conference in Paris in December 2015 limits global warming to less than 2°C compared with the preindustrial level. According to studies cited by the United Nations' International Panel on Climate Change (IPCC), this target, as agreed upon in Paris, will require an end to fossil energy well before 2050, as discussed above.

The question becomes, how can we get the carbon Footprint to zero without shifting the burden onto other Footprint components? If we assume a linear reduction, this would require a reduction of 600 million global hectares every year. Or about 0.1 global hectare per person per year. Since lower Ecological Footprints are harder to shrink, annual per person reductions of 0.3 global hectares in high-Footprint countries may be a more reasonable assumption.

Earth Overshoot Day uses calendar days to measure Footprint and the changes necessary to reach the 2015 climate conference target. One reason is that it makes results understandable. Very few relate to 2°C, ppm, or tons of carbon (or did you mean CO_2 which is 3.7 times more?). But even elementary school kids understand number of planets or dates. It is graphical to report that, in 2019, humanity had used up the annual budget of nature by July 29th. And it makes sense that it was safer (but not safe) to have Earth Overshoot Day on September 23rd in the year 2000 rather than on July 29th. To reach the Paris 2030 milestone of a 30 percent emissions reduction compared to 2005, this would require moving the date another 7 days into the future, every year. If we move the date at that speed until 2042, we'd be back at one planet. That's is why we call our campaign to get out of ecological overshoot #MoveTheDate.

Global Footprint Network estimated what it would take to move the dates using sources from other groups, like Drawdown.org. We came to the conclusion that cutting carbon emissions would move the date of Earth Overshoot Day by 90 days—nearly three months. With Schneider Electric, we estimated how much the date shifted if the entire world adopted solutions which Schneider offers for housing stock and electric grids. Such retrofitting alone would move the date 21 days, without losing, if not increasing, comfort.[9]

We can also reach higher, setting the goal E. O. Wilson suggested, to half a planet for humanity. If we want to reach this goal by 2100, humanity would need to continue shaving off 1.5% of its Ecological Footprint every year after 2050. For this scenario, the role of population is dramatic. If we opt for much smaller families, our chances of achieving such a goal are far higher.

And remember, 2050 is not that far away, and 4.3 billion people alive today will most likely also be alive in 2050. All those born after 1985 will still be under 65 years old by then.

How to achieve these reductions was discussed in Chapter 4. Technology, if applied well, can be of enormous help reaching these goals. But if not managed well, as explained by Jevon's rebound effect,[10] the challenge could be even harder.

Of course, we can imagine any number of alternative scenarios besides the two outlined in this chapter. Whatever we choose, we will not be able to reach the core goal—ending overshoot—on a straight path but will face crises along the way, even if we do our best to prevent them.

To eliminate overshoot means to close the gap between humanity's Footprint and the planet's biocapacity. For that to happen, the world's community of peoples will have to agree in principle on the degree by which to reduce Footprint as well as on a mechanism for allocating demands on our natural capital among individuals and groups of people. If we do not reach an international agreement, our world will become unpredictable even faster. Or default solutions will emerge—the retreat to nation-states, with little international cooperation.

Whatever path will be taken, for countries and cities this means that they should prepare much more aggressively to protect their quality of life when most economies become resource poor.

Possible allocation mechanisms propose either fixed Footprint shares for individuals or groups or argument and resolution using *consumption rights*, which essentially parallel the trade in pollution rights we know from current climate strategy. Consumption rights could be granted to individuals, countries, or regions. Any acceptable global strategy will have to take into consideration ethical, economic, and ecological concepts, and none will be able to resolve all contradictions.

In the text that follows, three possibilities for allocating Ecological Footprint reductions are introduced. The purpose is to stimulate discussion. Many more ways than these three are conceivable. Different regions may choose different approaches. It probably won't go smoothly. But maybe there will be a global agreement on determining rights of access to biocapacity. Such rights could be determined relative to a region's supply of biocapacity or relative to the size of a region's population. Allocations could be fixed or flexible, depending on a region's development.

Allocation Approach #1—
Reduction of Footprint Relative to Historical Size

This approach starts at historical levels of consumption, similar to the mechanism the Kyoto Protocol developed for the reduction of greenhouse gas emissions. One main objection to this approach would be that such an allocation strategy privileges regions with historically high consumption levels and large populations. It also disadvantages those that have already begun to reduce their consumption and population.

Allocation Approach #2—
Footprint Allocation Based on the Size of a Region's Biocapacity

In this approach, the region's biocapacity determines how much a population is able to get, unless there are additional trading mechanisms in place that allow regions with reserves to trade with regions whose population wants more. Given that, over the last decades, many population hubs have expanded far beyond the resource capabilities of their region (e.g., Singapore, Hong Kong, UAE, South Korea), there might be a need for a transition strategy that allows those places to adjust.

Allocation Approach #3—
Footprint Distribution to Provide Equally to Every Person

This strategy would also require a transition strategy as the shifts in resource availability would for some countries be dramatic. We could imagine additional trade mechanisms to enable nation-states and regions with large Footprints to get extra support through trade from countries with biocapacity reserves. Obviously, for such trade to work, the owners of biocapacity would need to be compensated. This is different from climate schemes because, in contrast to the atmosphere, most productive areas belong to someone. The downside of this approach is a lack of incentive for regions to contain their populations. But this could be addressed by pegging the allocation against a population number pegged against a given time in history.

Similar proposals already exist in the climate debate. It is true that any approach would have to struggle with basic questions of what is fair. Each might lead to a political quagmire. History has shown that it is hardly possible to force major powers to send massive transfer payments to smaller countries. Military power imbalance alone makes it unlikely. Without a functioning world government with some clout, it is hard to imagine how such global redistribution could become reality. And every region would claim its own special needs. After all, exceptionalism is universal.

Negotiations that aim at reducing humanity's Footprint and to that goal deploy allocation mechanisms like those outlined above presuppose an unprecedented global willingness to cooperate. The challenges, complexity, and costs of such an undertaking are gigantic. They are surpassed, though, by the consequences for humanity and for the ecosystems should we fail at our endeavor.

Hence the proposal to look at the world with all its ecological flows and resources in mind instead of looking just at carbon dioxide. Such a perspective makes the self-interest of cities, regions, and countries much more visible. To wait and see how others react is becoming an act of self-destruction. Not acting means that we underprepare ourselves (while also keeping destroying humanity's home). We already know more about the future than we may like to know. Growing resource constraints have become a certainty. Cities and countries that fail to take proactive measures to prepare themselves will suffer. With the help of Footprint accounting, we can calculate how fast we need to adapt, and we can see how well we do, compared to the transformations elsewhere on Earth. Most communities and countries are dramatically underinvested in resource security. They risk experiencing shortages. Mayors and ministers don't seem to notice. They talk about how the world ought to be saved, but maybe they had better focus on saving their own city or country—such local action would offer the world as a whole much more benefit.

PART III

FOOTPRINT

Case Studies

CHAPTER 9

FOOTPRINT CALCULATIONS

Individuals, Cities, Countries, Products, and Companies

The accounting system of the Ecological Footprint is fully scalable. Hence it has a vast range of applications, from the production of a toothbrush to the resource consumption by all of humanity.[1] Whether we focus on individuals, cities, regions, countries, or the entire global population, national Footprint assessments are used as a reference point to ensure consistency when national governments keep track of production and trade, and their statistics therefore provide the most comprehensive reference point ensuring a coherent and consistent picture that meaningfully links to a global total. Calculating the Footprint for goods or services requires Life-Cycle Assessments (LCA) that get calibrated against these national assessments. A *Life-Cycle Assessment* identifies all resource and energy flows involved in the production, use, and disposal of a product. Footprint accounting lets us interpret the LCA information from a biocapacity perspective, expressed in global hectares.

An Individual's Footprint

In the calendar year 2018, 2.5 million unique visitors calculated their individual Footprint on footprintcalculator.org. It takes only a few minutes. You answer questions about your lifestyle and consumption habits as far as housing, mobility, food, and other consumption are concerned. Do you have access to a car? How often do you eat meat or meat products? What kind of house do you live in? A single or multi-family dwelling? How well is it insulated? The whole thing works like a little quiz. In the end, visitors are given a concrete result: the number of global hectares you require. The calculator also tells you how many Earths would be needed if all of humanity lived the way you do, and when Earth Overshoot Day would happen accordingly. It is not unusual that a North American or European city dweller with a well-paying job turns out to require four or five planets. The odd player may even be a little shocked at their results.

This is why the calculator sports an elaborate solutions section focusing on the already discussed four significant drivers, each one of which has a personal dimension as well:

1. How we build cities (where I live and how efficient my house is)
2. How we power ourselves (what your electricity mix is)
3. How we feed ourselves (your food waste and amount of animal products you consume)
4. How many we are (how big of a family you choose to have)

Global Footprint Network is planning to build out the solutions section and make if far more interactive.

The whole thing is set up as a rough approximation toward the reality of one's life, with blurriness in the details but reliability as for the general trend. For example, if someone lives in a big house, Footprint calculations assume that the house is furnished accordingly with furniture, carpets, and interior decorations. Pursuing one's calculations to the last decimal point—which is possible—would require more information and more time than makes sense in such a quiz. The players at their computers would likely soon lose interest.

In spite of all its simplifications, the Footprint calculator is a software program informed by Global Footprint Network's National Footprint and Biocapacity Accounts. These national numbers are the starting point. These national results get further analyzed through Multi-Regional Input-Output assessments to produce the already discussed Consumption–Land-Use Matrix that tell us how much of the national Ecological Footprint is occupied by which activity. The calculator then estimates, using your calculator answers, how far your consumption deviates from the established average. With this information, the calculator in essence constructs a personalized Consumption–Land-Use Matrix that details the Footprint distribution for you (including your total Footprint).

Using the Footprint calculator is typically an eye-opener: this much nature is required for the sum total of your demand on nature. At the end, the calculator shows options for intervention, but stays away from telling people what to do. This is deliberate as an imposed intervention will only increase people's reluctance. Therefore, the purpose of the calculator is to reveal the magnitude of the challenge, and emphasizes together, we desperately need to turn these trends around.

The game's implicit message simply reminds us that we have only this one planet and together must find ways to live within its constraints. Too easily, the calculator may trigger thoughts of guilt or feelings of imperfection. It might even evoke fears that it is calling for suffering and sacrifice. But the invitation we are really providing is to #MoveTheDate. That unless we can shift the collective Earth Overshoot Day, we are not winning. And pushing out this date gives us more space, more security, more opportunity for a thriving future. In other words, what we are really calling for is to use the information we have for foresight, which then allows us to channel innovation. What are your best ideas for a thriving future, given the one-planet context?

The next level of the calculator expansion will be a platform to enable users to share the one-planet solutions they feel passionate about.

A City Dweller's Footprint

The Footprint of a city dweller is calculated according to the same principle as any individual's Footprint. Again, the point of reference is the data set documenting the average demand of the country's citizen. We then estimate how far consumption of residents of the particular city deviate on average from the national average. Do they drive

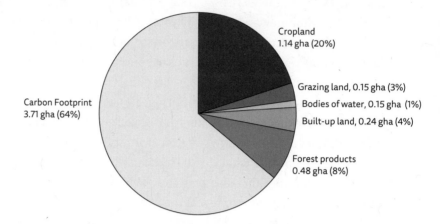

Figure 9.1. North-Rhine Westphalia Footprint per person by land-use category in 2012 in global hectares (gha). Credit: (based on study for NRW) Global Footprint Network as of July 2016.

longer distances to get to work? Do they live in bigger dwellings? Is it hotter or colder, needing more cooling or heating energy? Do they eat differently, more fish, more vegetables, less animal products than national average? How much do they earn compared to national average? And how expensive is it to live in this city? From the sum total of all these deviations, we can calculate the results for that city.

Recently, Global Footprint Network has been assessing the Ecological Footprint of six Portuguese cities, as well as of Nancy, a metropolitan area of about 450,000 inhabitants in the east of France. Not long ago, we assessed the Ecological Footprint of Germany's most populated province, North-Rhine Westphalia. Given that it is also one of the provinces with highest per person income, it also exceeds German's per person Footprint: 5.8 global hectares per person was the average we computed for North-Rhine Westphalia. Figure 9.1 depicts the surface area shares of the total. Not surprisingly, given industrial lifestyles and significant portion of coal-powered electricity, 3.7 global hectares of the total, or 64%, are taken up by the carbon Footprint alone. In other words, this would be the forest area needed to sequester enough of their emissions to not add new carbon to the atmosphere. The second-largest portion is cropland at 1.1 global hectares *per capita*, for food, animal feed, and clothes fibers. The third-largest is forest area for extracting wood and pulp.

This amount can now be compared to what is available. Humanity had about 1.63 global hectares per person on Earth. Germany has just about the same amount with 1.62 global hectares per person in 2016. North-Rhine Westphalia's biocapacity, on the other hand, amounted to around 1.1 global hectares per resident, partially due to its population density which, at 500 inhabitants per square kilometer (or 1,300 people per square mile), is more than twice that of Germany as a whole. The largest share of NRW's biocapacity is comprised of cropland and grazing land at a combined 0.64 global hectares per person (48% of land area of the province is agricultural land), followed by the forest category at 0.23 global hectares per person (for forest products and CO_2 sequestration). A similar area is occupied by built-up land and infrastructure contributing 0.20 global hectares per resident. North-Rhine Westphalia does not have any noteworthy amounts of productive bodies of water.

This means that North-Rhine Westphalia requires for its consumption currently 5.8/1.1 = 5.3 times more than what their own ecosystems can renew. And this does not even include any space needed by wild species.

How Reliable Are Such Results?

Robust results all depend on the available data. For many cities, it is possible to get information about at least basic consumption patterns. There are many commercial interests who want to know what and how much people buy, what cars they drive, and so on. These things are of interest to retailers, marketing departments, and advertisers. The unintended consequence is that we can also use these data sets to estimate the Footprint of these cities, sometimes even to the neighborhood level.

However, numbers from different statistical surveys usually don't fit well with each other; they are not consistent. This is why Footprint accounting always returns to national data as its reference point.

Methodologically, national data are much more reliable because they are collected across the world in comparable ways. So they remain the yardstick against which all lower-level units—regions, cities, or neighborhoods—are measured. That way we can properly compare data.

While financial data on consumption do not translate exactly into physical amounts, they do provide a useful approximation. Once we have a Consumption–Land-Use Matrix established for a particular city or region, then more detailed local data can be used to increase the resolution or estimate, over time, how this overall demand is changing.[2]

It increasingly matters that citizens of cities and regions as well as their representatives get a clear picture: How much nature do we need? Where do we get it from? How can we avoid unnecessary costs and dependencies? And not least, how can we deploy resources most wisely to secure a good standard of living?

For example, it may surprise people to learn that food makes up a considerable portion of any community's Footprint: often more than ¼, even in high-income cities. If a city draws on seasonal food supplies from the surrounding region, it can definitely shrink this portion. By contrast, supermarkets with their complex supply chains and highly processed food products are resource- and energy-intensive.

Of course one will always find exceptions and extremes in this regard. Under certain circumstances, the Footprint of a freshly harvested apple from Chile that is eaten in the US may not be larger than the Footprint of an apple grown locally. That is because apples are transported by sea, which is relatively efficient. And it also takes energy to store apples for a good portion of the year.

With the Footprint, it is important to look at the overall picture. And there again the rule of thumb proves valid: Food products that have to be transported long distances; that are preserved, cooled, or highly processed, have a larger Footprint than fresh, regional food. Animal-based food products always have a much higher Ecological Footprint as well.

From the Footprint perspective, energy and carbon dioxide are currently the main players in industrialized cities: they use up notably more than ½ of all required biocapacity. As discussed, carbon Footprints need to become zero soon around the world. It therefore makes sense for a city as a first step to lower its carbon Footprint. We now have hundreds of examples of cities approaching this task in pragmatic as well as systemic ways.[3] By lowering their carbon dioxide emissions, cities also save energy and hence cost.

Carbon dioxide makes up only half the problem. The essential question is not what's the biggest problem? Is it climate? Or food? Or maybe water? It is not a "beauty competition" between issues. Rather, it is the sum total of them. Using extra water, emitting extra CO_2, or requiring extra food all adds up to more biocapacity use. Therefore, it is essential to look at the entire metabolism of a city or region, to look at the sum total. The Footprint measures the areas that provide ecological services and supply a city with everything it needs. These areas are competing with each other, and scarcities multiply. Given overshoot and climate change, we are not only dealing with peak fossil fuel, we are dealing with peak everything. While it is absolutely fine for a city to start with its carbon Footprint, it must not stop there. In its own interest, the city has to go further.

To exclusively lower emissions does not offer a city any quick financial advantages. But if emissions reduction is seen as essential for the value-maintenance of its building infrastructure and is paired with energy savings as well as more conservative water consumption, local job creation, and regional investments in renewable energy, then the synergy of these efforts can create a program that can bring considerable benefits to that city.

Only with the overall situation in mind can we succeed in making meaningful investment decisions for a community's future development while assuring the sustainability of its infrastructure. And only by looking at the larger picture will we be able to answer questions such as: What building material—brick, concrete, or wood—is more eco-friendly and hence also more economical in the long run? What do architects have to take into account in any particular climatic conditions? Should public transport in our city run on gas or electricity; or be replaced with public scooters and bicycle lanes?

City planners know that a compact city is a resource-efficient city. Still, political reality often does not support high-density development. Even though it has long become obvious that one's house in the suburbs or countryside will lose much of its value over the next decades because it necessitates long and therefore expensive travel—for many such housing remains their most cherished dream. Sparsely settled areas, especially in parts of the United States, Australia, or Canada, are becoming social traps as gas prices rise. One day the drive to work is no longer worth it. At that point, an impoverished existence is all that is left for many rural residents.

Settlement structures are extraordinarily long-lasting. Acting early is therefore hugely important. The Footprint helps by translating complexity into a simple number. Footprint accounting gives local politicians and city planners a tool to bolster their argument in favor of a compact city and to effectively share it with the residents. Over the years, the Canadian city of Calgary, for example, developed suburb after suburb because of urban sprawl. Only after a Footprint analysis of the city and the subsequent discussion did Calgary pass a moratorium on sacrificing new green space.[4]

Cities have been active, now also with many city organizations that are actively pursuing sustainability performance, including ICLEI, C40 Cities, or Ecocity Builders.[5] David Thorpe has also collected many examples for his book on "One Planet" cities.[6]

A Country's Footprint

The Footprint of a country indicates how much biocapacity is required to supply the goods and services used by the residents of that country. It also includes the capacity needed to neutralize the waste they leave behind. Waste is a component of a national Footprint because biocapacity is required to manage it and eventually absorb it. Any goods a country exports are calculated as part of the Footprint of the nation that consumes them.

Global Footprint Network's most common Ecological Footprint numbers are consumption based, meaning they show how much it takes to support what all residents consume. It is also possible to look at countries from the production side—how much the economy of a country directly extracts from nature. Globally, production and consumption are identical as everything consumed on the planet is produced on the planet. But country-by-country production and consumption vary because of trade.

So much for the principle. But how does the calculation work? How can we measure and compare car tires, door handles, pork sides, insurance policies, shoes, T-shirts, and crayons? Thanks to the United Nations, fairly consistent statistics exist for all countries. They show not only how much a given country's industry, agriculture, and forestry produce but also how many goods that country exports and imports. On the basis of these statistics, Footprint accounting can calculate how much biocapacity is consumed inside a country. How much rice "disappears" in Canada? How much wood? How much energy? Dividing these amounts by the number of residents gives us Canadian *per capita* consumption. The result is a relatively stable estimate, based on as many consistent UN data sets available. Currently, this means per country and year, Global Footprint Network National Footprint and Biocapacity Accounts use up to 15,000 data points from the United Nations statistics.[7] UN statistics may not be the most precise ones, but they are most widely accepted.

Footprint methodology at the national level is well developed, and much focus has been put on developing those accounts as they are foundational for any assessment. The task of maintaining, improving, and standardizing the National Footprint and Biocapacity Accounts has been transferred to a new independent organization as the work scope has outgrown Global Footprint Network and needs an independent home. This new home will be supported by a global network of academic partners.[8]

Below National Footprint and Biocapacity Accounts, a deeper layer of analysis can be produced: Consumption–Land-Use Matrix (CLUM). Based on financial flow data between economies' sectors,

Global hectares per person	Cropland	Grazing land	Forest products	Fishing grounds	Built-up land	Carbon	Total
Food	0.37	0.10	0.04	0.02	0.00	0.16	**0.69**
Housing	0.09	0.02	0.57	0.01	0.01	1.34	**2.02**
Personal transportation	0.04	0.01	0.10	0.00	0.01	0.74	**0.89**
Goods	0.09	0.04	0.18	0.01	0.00	0.40	**0.72**
Services	0.08	0.02	0.16	0.01	0.00	0.53	**0.80**
Total	**0.66**	**0.19**	**1.05**	**0.05**	**0.02**	**3.17**	**5.13**

Figure 9.2. High-level Consumption–Land-Use Matrix (CLUM) for Slovenia in global hectares per person; data year 2014. Does not add up exactly due to rounding of components. Credit: Global Footprint Network MRIO.

it is possible to construct an Input-Output assessment that allocates overall demand to final consumption categories. The data Global Footprint Network uses to have a consistent assessment across the world is from the Global Trade Analysis Project (GTAP) database hosted at Purdue University.[9] It enables Global Footprint Network to assess the CLUM for about 80 countries, for three distinct years: 2004, 2007, and 2011. GTAP allows to break down consumption into 37 categories.

Figure 9.2 shows a summary example for Slovenia. Vertically, the table lists the main components of consumption: food, housing, mobility, goods, and services. In theory, it is possible to create numerous additional subcategories. But in practice, any further differentiation depends on the quality of data sets and, not least, on the budget of any given project.

Horizontally, the table shows the six different area types that Footprint accounting recognizes: cropland, grazing land, forest products, fishing grounds, built-up land, as well as carbon Footprint (or the forest for the absorption of carbon dioxide from fossil fuel burning). The numbers at the intersections of vertical and horizontal categories show how much biologically productive area is needed for nature to provide the respective service, such as producing food or moving about. The measurement unit for all numbers is *per capita* global hectares.[10] Overall, the table (Figure 9.2 and 9.3) shows the patterns of consumption of a resident of Slovenia, on average. Their Footprint amounts to 5.13 global hectares per person.

National Footprint calculations are currently created for every country with a comprehensive United Nations data set, that is, for all 153 countries with populations of more than one million people, plus an additional 215 countries with smaller populations. There are more small nations, but they may not have complete enough data sets to warrant an assessment. The results are assigned quality data scores of a number and a letter; the score indicates how complete the data sets are that the results are based on. (The top score is 3A. It is awarded for accounts with complete data sets for each consecutive year.)[11]

Ranked as the countries with the largest Footprints are Qatar, Luxembourg, United Arab Emirates, Mongolia, Bahrain, and United States of America with 7.5 to 14.2 global hectares per person. At the other end are the countries with the lowest *per capita* Footprints, which are currently Haiti, Burundi, Timor-Leste, and Eritrea with 0.5 to 0.7 global hectares per person. So much for the demand side.

The supply of biocapacity is also estimated using UN data. That in return may be generated partially using satellite imagery, among other sources. Biocapacity assessments show how much biologically productive area is available to a country. These data get processed and converted into the "currency" of global hectares.[12] A comparison of supply and demand allows us to see whether a country is a biocapacity creditor or debtor.[13]

Meanwhile, Global Footprint Network has encouraged or participated cooperatively to assess the validity of the National Footprint and Biocapacity Accounts with national governments. This has led to review by the following governments: Belgium, Ecuador, France (three times), Finland (for its forest Footprint), Japan, Switzerland (twice), United Arab Emirates (several times), Luxembourg, Germany, Indonesia, and the Philippines, as well as the European Union.[14] The governments of Montenegro, Slovenia, Costa Rica, Latvia, Argentina, Wales, and Scotland also make official use of the Footprint methodology.

Figure 9.3. Detailed Consumption–Land-Use Matrix (CLUM) for Slovenia in global hectares per person; data year 2016. Does not add up exactly due to rounding of components. Credit: National Footprint and Biocapacity Accounts MRIO.

	Cropland	Grazing land
Short-term consumption paid for by household		
Food	**0.36**	**0.08**
Bread and cereals	0.10	0.00
Meat	0.06	0.05
Fish and seafood	0.01	0.00
Dairy	0.03	0.02
Vegetables, fruit, nuts	0.15	0.01
Other food	0.02	0.00
Non-alcoholic beverages	0.00	0.00
Alcoholic beverages	0.01	0.00
Housing	**0.02**	**0.01**
Actual rentals for housing	0.00	0.00
Imputed rentals for housing	0.01	0.00
Maintenance and repair of the dwelling	0.00	0.00
Water supply and miscellaneous services relating to the dwelling	0.00	0.00
Electricity, gas, and other fuels	0.01	0.00
Services for household maintenance	0.00	0.00
Personal Transportation	**0.03**	**0.01**
Purchase of vehicles	0.01	0.00
Operation of personal transport equipment	0.02	0.01
Transport services	0.00	0.00
Goods	**0.08**	**0.04**
Clothing	0.03	0.02
Footwear	0.01	0.01
Furniture and furnishings, carpets and other floor coverings	0.00	0.00
Household textiles	0.00	0.00
Household appliances	0.00	0.00
Glassware, tableware, and household utensils	0.00	0.00
Tools and equipment for house and garden	0.00	0.00
Medical products, appliances, and equipment	0.00	0.00
Telephone and telefax equipment	0.00	0.00
Audio-visual, photographic, and information processing equipment	0.00	0.00
Other major durables for recreation and culture	0.00	0.00
Other recreational items and equipment, gardens, and pets	0.00	0.00
Newspapers, books, and stationery	0.00	0.00
Goods for household maintenance	0.00	0.00
Tobacco	0.02	0.00

Forest products	Fishing grounds	Built-up land	Carbon	Total
0.03	0.02	0.00	0.16	0.66
0.00	0.00	0.00	0.01	0.10
0.01	0.00	0.00	0.04	0.16
0.00	0.01	0.00	0.00	0.02
0.01	0.00	0.00	0.04	0.09
0.01	0.00	0.00	0.03	0.20
0.00	0.00	0.00	0.01	0.03
0.00	0.00	0.00	0.01	0.03
0.00	0.00	0.00	0.01	0.02
0.43	0.00	0.00	0.90	1.36
0.15	0.00	0.00	0.02	0.18
0.02	0.00	0.00	0.04	0.07
0.00	0.00	0.00	0.01	0.01
0.02	0.00	0.00	0.04	0.07
0.23	0.00	0.00	0.79	1.04
0.00	0.00	0.00	0.01	0.01
0.08	0.00	0.00	0.64	0.77
0.01	0.00	0.00	0.06	0.08
0.06	0.00	0.00	0.53	0.62
0.01	0.00	0.00	0.05	0.06
0.16	0.01	0.00	0.31	0.60
0.03	0.00	0.00	0.10	0.19
0.00	0.00	0.00	0.02	0.03
0.08	0.00	0.00	0.02	0.10
0.00	0.00	0.00	0.00	0.01
0.00	0.00	0.00	0.01	0.02
0.00	0.00	0.00	0.00	0.01
0.00	0.00	0.00	0.01	0.01
0.00	0.00	0.00	0.02	0.03
0.00	0.00	0.00	0.00	0.00
0.00	0.00	0.00	0.01	0.02
0.00	0.00	0.00	0.00	0.00
0.01	0.00	0.00	0.03	0.04
0.02	0.00	0.00	0.03	0.06
0.00	0.00	0.00	0.01	0.01
0.01	0.00	0.00	0.04	0.07

Figure 9.3. (cont'd.) Detailed Consumption–Land-Use Matrix (CLUM) for Slovenia in global hectares per person; data year 2016. Does not add up exactly due to rounding of components. Credit: National Footprint and Biocapacity Accounts MRIO.

	Cropland	Grazing land
Services	**0.03**	**0.01**
Out-patient services	0.00	0.00
Hospital services	0.00	0.00
Postal services	0.00	0.00
Telephone and telefax services	0.00	0.00
Recreational and cultural services	0.00	0.00
Package holidays	0.00	0.00
Pre-primary and primary education	0.00	0.00
Catering services	0.00	0.00
Accommodation services	0.00	0.00
Personal care	0.00	0.00
Personal effects n. e. c.	0.00	0.00
Social protection	0.00	0.00
Insurance	0.00	0.00
Financial services n. e. c.	0.01	0.00
Other services n. e. c.	0.00	0.00
Household Sub-Total	0.53	0.14
Short-term consumption paid for by government	0.06	0.01
Long-term infrastructure construction (Gross Fixed Capital Formation)	0.07	0.04
Total	0.66	0.19

The Footprint of Products and Services

Let's take a toothbrush. How large is its Footprint? Here, national statistics won't get us very far. We rather need to ask, What is the toothbrush made of? How many resources and how much energy were consumed up to the moment when we put that toothbrush into our shopping cart in the supermarket? In fact, we have to think even further ahead and consider what will happen once that toothbrush lands in our garbage. How much biocapacity does it cost to dispose of it?

Fortunately, we now have the relevant data for many products and services. A *Life-Cycle Assessment (LCA)* systematically examines the environmental effects of a product from the moment natural

Forest products	Fishing grounds	Built-up land	Carbon	Total
0.06	0.00	0.00	0.20	0.31
0.00	0.00	0.00	0.01	0.02
0.00	0.00	0.00	0.00	0.01
0.00	0.00	0.00	0.00	0.00
0.01	0.00	0.00	0.02	0.04
0.01	0.00	0.00	0.03	0.04
0.00	0.00	0.00	0.02	0.02
0.00	0.00	0.00	0.00	0.00
0.00	0.00	0.00	0.01	0.01
0.00	0.00	0.00	0.00	0.00
0.00	0.00	0.00	0.01	0.02
0.00	0.00	0.00	0.01	0.02
0.00	0.00	0.00	0.01	0.01
0.00	0.00	0.00	0.01	0.02
0.01	0.00	0.00	0.04	0.07
0.01	0.00	0.00	0.02	0.03
0.76	0.03	0.01	2.21	3.70
0.11	0.00	0.00	0.40	0.59
0.16	0.00	0.01	0.56	0.84
1.03	0.04	0.02	3.17	5.13

resources are first used to create it to the disposal of the product in the end: from cradle to grave. A whole scientific field now focuses on various methods and their respective international standards.[15]

A Life-Cycle Assessment functions like a baking recipe in reverse. You look at the finished cake and ask, What did it take to bake such a cake? Only, a Life-Cycle Assessment is more thorough. It is not satisfied with just the ingredient of, say, one kilogram of flour but asks, Where does the flour come from? How was it milled? How much energy went into the milling? How many resources vanished in the processing of the grain? In this manner, a Life-Cycle Assessment traces the entire life cycle of all ingredients through all stages of production.

Figure 9.4. Criteria for Data Quality Scores Used in Publication of Country Results. Each country in the National Footprint and Biocapacity Accounts 2018 Edition is given a quality score comprised of two elements, time series score [1–3] and latest year score [A–D]. Credit: Global Footprint Network.

	No component of biocapacity or Ecological Footprint is unreliable or unlikely for any year.
3A **3B**	No component of biocapacity or Ecological Footprint is unreliable or unlikely for the latest data year. Some individual components of the Ecological Footprint or biocapacity are unlikely in the latest data year. The total Ecological Footprint and biocapacity time series results are not significantly affected by unlikely data.
3C	No component of biocapacity or Ecological Footprint is unreliable or unlikely for the years prior to the latest data year. Some individual components of the Ecological Footprint or biocapacity are unlikely in the latest year. Total Ecological Footprint and biocapacity values are unlikely or unreliable in the most recent data year, but the ability to ascribe creditor/debtor status is unaffected in latest year.
3D	No component of biocapacity or Ecological Footprint is unreliable or unlikely for the years prior to the latest data year. Some components of the Ecological Footprint or biocapacity are very unlikely in the latest year. Ecological Footprint and biocapacity results in the latest year are significantly impacted by the unlikely or unreliable values, making them unusable.
2A	Ecological Footprint or biocapacity component time series have results that are very unreliable or very unlikely, except in the latest data year. The total Ecological Footprint and biocapacity time series results are not significantly affected by unlikely data. No Ecological Footprint and biocapacity results in the latest year are significantly affected by unlikely data.
2B	Ecological Footprint or biocapacity component time series have results that are very unreliable or very unlikely, including the latest year. The total Ecological Footprint and biocapacity time series results are not significantly affected by unlikely data.

2C   Total Ecological Footprint or biocapacity time series and component Ecological Footprint and biocapacity time series results are unreliable or unlikely, especially in the latest year. The total Ecological Footprint and biocapacity time series results are not significantly affected by unlikely data. The unlikely or unreliable values have most likely not impacted the creditor/debtor status in the latest year.

2D   Total Ecological Footprint or biocapacity time series and component Ecological Footprint and biocapacity time series results are unreliable or unlikely, especially in the latest year. The total Ecological Footprint and biocapacity time series results are not significantly affected by unlikely data. Ecological Footprint and biocapacity results in the latest year are significantly impacted by the unlikely or unreliable values, making them unusable.

1A   Several components of the Ecological Footprint or biocapacity are very unreliable or unlikely, except the latest year. The Ecological Footprint and biocapacity time series results are significantly affected by unlikely data, and are unusable. No Ecological Footprint and biocapacity results in the latest year are significantly affected by unlikely data.

1B   Several components of the Ecological Footprint or biocapacity are very unreliable or unlikely, except the latest year. The Ecological Footprint and biocapacity time series results are significantly affected by unlikely data, and are unusable. The total Ecological Footprint and biocapacity results in the latest year are not significantly affected by unlikely data.

1C   Several components of the Ecological Footprint or biocapacity are very unreliable or unlikely. The Ecological Footprint and biocapacity time series results are significantly affected by unlikely data, and are unusable. The unlikely or unreliable values have not impacted the creditor/debtor status.

1D   There is too much unreliable or unlikely data to make any conclusions about the timeline or latest year of this country.

Note: Through further nation-specific research, preferably in collaboration with researchers from those countries (particularly from government agencies), it is possible that the Data Quality score (i.e., the quality of the results) can be improved. Improved data sets, methodological improvements in the National Footprint and Biocapacity Accounts, and better data cleaning processes have also helped to increase the Data Score of some country results in past editions, as is likely in the future.

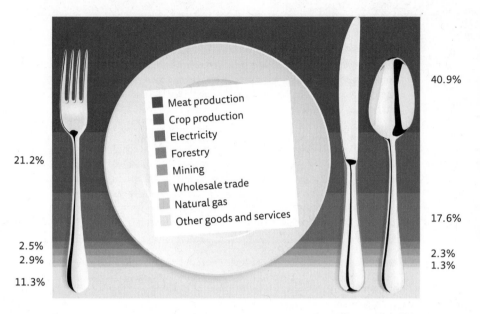

40.9%

21.2%

17.6%

2.5%

2.9%

2.3%

1.3%

11.3%

Figure 9.5. The Footprint of a restaurant meal. Credit: EPA Victoria/SEI.

Luckily, vast databases exist with precalculated factors for most aspects of consumption. For instance, if you are interested in mobility, you may want to play with the freely available mobitool spreadsheet, and learn some German in the meantime.[16]

Here is an example from past studies: just one average restaurant meal in the Australian state of Victoria requires 61 global square meters for one year. The largest impact comes from the area needed to grow the food in the first place, with meat taking the top position. But all the other services to serve that meal at the table must be factored in as well, such as the energy for cooking.

Let's take another example: a piece of paper. Paper production has become notably more efficient over the centuries, using less wood fiber, less water, and less energy per kilogram of paper. These days, paper can also be recycled or rather downcycled, since each cycle lowers the paper's quality. How often paper is recycled also enters into its eco-balance sheet. It's only a short step from one piece of paper's eco-balance sheet to the Footprint of an entire newspaper. Typical newsprint has a Footprint about the same size as the paper itself (depending on percentage of recycled content—in the paper, not the news). Take the *New York Times* Sunday edition and spread it out, sheet by sheet. This will give you the area needed (occupied for one year) to provide this newsprint.

Now we can ask, What would change if for pulp we were to use rice straw instead of wood fiber? Because rice straw is a by-product, we could achieve an ecological synergy effect. Of course a Life-Cycle Assessment of rice straw allows us to determine a Footprint, too. Comparing the two Footprint numbers will tell us whether switching from wood fiber to rice straw would benefit the biocapacity bottom line.

Footprint as an aggregate indicator irons out the intricate wrinkles of Life-Cycle Assessments. It helps interpret the massive outputs such assessments generate. Translating them into biocapacity synthesizes them into just one single number. Sometimes, though, it is not only biocapacity that is relevant. For instance, when toxicity is involved. In such situations, we need to make sure to look not only at the Ecological Footprint. A product designer will also want to consider what the product will cost or whether materials are involved that endanger the health of humans or other living beings (for example, are toxins used that will accumulate in nature?).

Footprint can be deployed with a wide range of applications. For example, we may want to know whether agrofuels made from sugar cane, corn, or palm oil need more, or less, biologically productive area than conventional fuel made from crude oil. The question can be answered. For instance, a VW Jetta with an annual mileage of 20,000 kilometers (or 12,000 miles) might take about 1.6 global hectares if run on gasoline, and about 2.8 global hectares if it was fueled with B100 biodiesel (based on US soy). Three quarters of that Footprint would be cropland. E85 from US ethanol would have an even worse balance (possibly up to 3.7 global hectares), while E85 based on Brazilian ethanol would be about the same as gasoline, with the difference that about half of this Footprint would be from cropland, and the other half from its carbon Footprint. A Toyota Prius, being more efficient than the Jetta, would require 0.85 global hectares of biocapacity to drive the same distance per year on conventional gasoline.[17]

It might surprise us that the consumption of conventional fuel typically requires markedly fewer global hectares than current agro-fuels. For this calculation, we have of course to factor in two land types: the land type needed for absorbing the CO_2 of the fossil fuel used in the production of the agrofuel as well as the cropland needed for growing energy crops.

Let us close with another example where biocapacity, while a constraint, is not the limiting factor. There are even tighter other environmental constraints in the case of rain forests that are clear-cut to make room for palm oil plantations. Rain forests absorb and retain a lot of carbon dioxide and are home to an extraordinary diversity of species, among them the highly endangered orangutan. Too much demand even in this one component may have long-term consequences that show up way before global overshoot, or even as other parts of the country's ecosystems may be "underused." Living within biocapacity constraints is a necessary but not always sufficient condition to maintain the ecological capital intact. We also know for a fact that any global overshoot necessarily implies many areas with local overshoot.

The Footprint of Companies and Branches of Industry

Companies, too, have to ask themselves whether they are living beyond their means or within the realm of the possible. Possible is of course a relative term, and it all depends on how far we look. Take cars, for example. What are all the aspects we need to consider? Do we focus merely on the consumption of resources by the car factory? Or do we include the many steps occurring before the final assembly of the car, such as the production of the steel and plastics for the car, and of the rubber for the tires? Then there is the life of the car when it is being used. Fuel consumption is the obvious demand, but cars also require infrastructure, namely roads and other traffic facilities such as bridges. They, too, consume resources to build and maintain.

The Footprint invites us to take a comprehensive view, to include everything it takes to provide for a product or a service. This is helpful for both companies as well as their clients to understand. But to take it a step up: What makes companies one-planet compatible, and how can we identify them? We would argue that these are companies that have long-term success built into them because they improve people's well-being while helping humanity to move out of ecological overshoot. Not only are one-planet companies compatible with the world that the UN Sustainable Development Goals (SDGs) and the Paris Agreement call for. More importantly, they also have a baked-in economic advantage: on average these companies aligned with this growing need to operate within the regenerative means of our planet are much more likely to be economically successful in the long run than those companies that are incompatible with one-planet prosperity and will inevitable face a shrinking demand.

But let's look here first at some examples of businesses that have applied the Footprint.

The first one is one of the earliest examples of Footprint applications by companies. It stems from the GPT Group,[18] a multinational real estate company with annual sales of several billion dollars. The company owns and manages shopping malls across Australia. The story begins with an environmental officer in the company who calculated that every square meter of sales floor required about 2,000 square meters of biologically productive area—a number that does not even factor in the products sold there, but reflects only what is needed to keep the sales floor operational: to build it in the first place; to maintain it; cool, light, and clean it, and so on. Rather than firing the environmental officer for exposing them, GPT management saw in the number an opportunity to increase the company's profitability. Clearly, the number reflected an extraordinary amount of waste and inefficiency. They began to look for a standardized method which would help them aim for a 20% reduction target in their properties' environmental impacts. Specifically, the company wanted to compare the environmental impacts of different architectural and interior design options they might use in a restructuring.

To meet their target, GPT Group teamed up with Global Footprint Network, and together they developed a software program that allows future tenants of GPT properties to calculate their Footprint. Detailed data for raw materials in various retail sectors, such as fashion stores, restaurants, or grocery stores, enabled Global Footprint Network to develop simple questionnaires to measure the Footprint of differently designed retail spaces. The program's goal is to persuade tenants to choose interior designs with small Footprints. Such designs will pay off financially, too, for both landlord and tenant.

The calculation principles developed for GPT serve as a useful base from which to compare a range of environmental impacts. Store fittings are closely examined for their use of materials and their generation of waste. In the process, cost savings and environmental protection potentials are identified. In addition, the software program indicates whether a business is making progress toward its sustainability targets by reducing its Footprint.

With this approach, GPT Group won early on a tender for a new shopping mall in spite of higher construction costs upfront. What determined the outcome was their ability to prove that their realistic sustainability targets would protect the mall's value over time. With the help of Footprint accounting, GPT was able to demonstrate that their blueprint required upwards of 20% fewer resources than those of their competitors. When the mall was built, calculations had even surpassed those promises. Despite the difficult economic climate created by the financial and economic crisis in 2008, GPT Group remained committed to Footprint as core indicator.[19]

That wonderful environmental officer who was not fired is Caroline Noller. She completed a PhD on the topic and started The Footprint Company which continues to lead such efforts around the world. For example, she helped with the renewal of the core of the Australian National University in Canberra. Caroline reported: "We implemented the Ecological Footprint as our sustainability improvement framework. The team had previous experience with the process, some going all the way back to the initial project with GPT. They pushed for a performance of 1.0 planets or better. I'm exceptionally proud of how this has come together, with every aspect of the project and every stakeholder being held accountable to halving the impact of their elements. These were the measurable outcomes:

- 40% operating energy reduction with a 10% contribution from embedded renewables
- On track for a 43% reduction in embodied carbon footprint of materials—with some really astounding innovations which have saved time and money
- A 30% reduction in transport impact through the mixed-use approach matched with interconnectedness with the city's cycle and bus infrastructure
- A 30%+ increase in space utilisation (increase in number of people using the same space) through shared / flexible space and programming (this was particularly challenging!!)
- Improving site bio-capacity through a precinct scale climate adaptive design."

"Most importantly," Caroline summarized, "the project outcomes were achieved in an extremely constrained commercial environment," a beautiful euphemism for achieving all this at no to low cost.

Let's look at another example: Schneider Electric. Global Footprint Network has developed a collaboration with Schneider Electric, a leading, originally French-based company driving digitalization and energy efficiency. Its strategy is geared toward reducing ever more aggressively CO_2, cutting costs and increasing resilience for its clients. Therefore, Global Footprint Network researchers teamed up with engineers from Schneider Electric to estimate what is possible right now using their technology. The goal was to find examples that are able to #MoveTheDate—to push the Earth Overshoot Day back. We focused on options for reducing demand as well as decarbonizing energy generation.

How much carbon reduction these options can generate depends on both technology (how inherently efficient the appliances are) and on how people use the technology. In order not to exaggerate, we asked ourselves, even without any shift in human habits, how many days could Earth Overshoot Day be moved using current off-the-shelf commercial technologies for buildings, industrial processes, and electricity production?

Since we wanted to analyze what is possible already today, the assessment focused on retrofitting existing buildings and industrial processes. On the energy side, we estimated current decarbonization opportunities of the electricity systems, based on current grid limitations. We found that these opportunities can move the date by 15 days. The energy retrofit and the decarbonization of electricity combined would move the date by over 21 days.[20] This is a conservative estimate as it is based on Schneider Electric's tested offerings. In addition, emerging technologies as well as technologies outside of Schneider Electric's realm may well exist to make those sectors even more efficient and move the date even further. On top of that, shifting people's resource use habits also holds a large potential.

Since much infrastructure is already built, and a big portion of it will last for decades to come, decarbonization goals will only be reached if we find more and better ways to retrofit the existing built environment. To succeed, retrofitting is therefore crucial—but often forgotten.

Retrofit is ultimately the art to use what already exists as effectively as possible. Typically, in Europe, buildings use about 40% of all the energy society consumes, generating almost the same percentage of the overall greenhouse gas (GHG) emissions. Here are some results from Schneider Electric's recent retrofit projects: Installing active energy efficiency systems in five different building types resulted in energy savings ranging from 22% in an apartment in Vaux-sur-Seine near Paris built in 2010 to 37% at a three-star hotel in Nice built in 1896 to 56% at a one-story primary school in Grenoble, France.

The improvements included addressing intermittent occupancy levels (such as putting controls on idle mode when hotel rooms are unoccupied); using CO_2 sensors to better control temperature and air quality; optimizing heating by occupancy level; controlling ventilation with CO_2 sensors; opening and closing blinds to leverage the sun for free natural light and heat, or to keep heat out. Applying these results to buildings across Europe, Schneider concluded that 40% of the building sector's total final energy consumption could be saved with active energy efficiency, representing a 16% reduction of Europe's overall energy bill.[21]

In Dallas County, the ninth-largest county in the US, government spent $600,000 on 54 buildings for improvements including mechanical system upgrades, water conservation controls and fixtures, and lighting with motion sensors. The project is expected to reduce utility bills by 31%, ultimately saving $73 million over 10 years. The county expects to reduce CO_2 emissions by more than 500,000 tons over that time period, which equates to removing nearly 8,500 cars from the road for these ten years, or planting 125,000 trees.[22]

This makes obvious why Global Footprint Network has partnered with Schneider Electric. Companies whose strategy is aligned with the growing need to live within the means of our one planet are much more likely to succeed in the long run compared to companies that are incompatible with one-planet prosperity and will inevitably face shrinking demand.

Global Footprint Network is proud to promote such companies because they are the most critical engines for the needed sustainability transformation. The strategic thinking and business focus Schneider Electric displays, i.e., an approach informed by the characteristics of our physical reality, is what we want to see become the norm, not remain an exception.

An older example goes back to a visit of a delegation of the Confederation of Indian Industry at Global Footprint Network in Oakland, California, in 2007. "We currently have an annual growth rate of nine percent," they said (this was before the economic crisis). "We want to get to 10% over the course of the next twenty years. But we don't want to cut the branch we are sitting on." The economic crisis of 2008 and slowdown of 2011–2013 has put a slight dent into economic expansion in India. Still from 2002 to 2017, India's economy expanded, on average more than 11% every year. Those branches are already creaking. Even though biocapacity has been boosted significantly through intensified agriculture, India now is using two and a half Indias every year, up from two in 2000. Also, since they visited

Global Footprint Network in 2007 until 2016, India's total Ecological Footprint has grown by 34%.

The Indian visitors had very specific questions. They were looking for a tool to measure the pressure on their ecosystems. With such a tool, they planned to compare different sectors, such as steel, paper, or energy production. They also wanted to assess and compare the consumption of natural resources by various factories or power plants within each sector. In short, their visit was about efficiently managing India's biocapacity.

The outcome of this cooperation between the Confederation of Indian Industry and Global Footprint Network led to a description of India's Footprint in relation to the country's biocapacity. While people in India had an average Footprint of only 0.86 global hectares in 2000, it expanded to 1.17 global hectares in 2016, which is still low compared to the world average. Obviously, individual Footprint numbers vary even more across society. Many members of India's growing urban middle class live with material consumption levels comparable to those of Europeans. But India's *per capita* biocapacity amounts to only 0.43 global hectares, merely a quarter of the world average.

It's only natural to ask whether this is fair. Obviously it is not.

Global Footprint Network was commissioned by the Confederation of Indian Industry to summarize these findings in a report. In spite of its troubling message, the report was kindly received and ignored. Was it too soft to be shocking and to become actionable? Or was it too troublesome to be embraced?

To translate our own findings into action, Global Footprint Network found opportunities to engage with some of India's most innovative development organizations, exploring with them the potential of sustainable development-driven metrics. Global Footprint Network calls this approach *Sustainable Development Return on Investment* (SDRoI). One of the organizations testing it out is IDE-India, which was founded by Amitabha Sadangi, a far-sighted, brilliant social entrepreneur. IDE-India enables local markets to provide basic, affordable technologies and products for sustainable agriculture. Examples are treadle pumps for garden irrigation, simple kits to help make fertilizer from compost, and organic insecticides. With its sustainable technologies, IDE-India has helped three million people escape extreme poverty at a one-time cost of fewer than $30 per person.[23]

As a result of these technologies, the families supported by IDE-India have become net producers and are able to turn their own wheel of progress. IDE-India operates from the core belief that resource security—in this case, food security—is an essential precondition for successful development. Global Footprint Network worked with IDE-India and the villages it supports to document the progress in human development that is driven by increased resource security and to facilitate comparisons with other development efforts.

This project builds on the United Nations Human Development Index (HDI)—Ecological Footprint diagram that contrasts human development performance against resource security (see Figure 5.4 in Chapter 5). The difference is that, with this village level initiative, we assessed the United Nations Human Development Index (HDI) for the village (which has not been done before as far as we know). Because if we cannot measure how a project impacts the local HDI, we cannot effectively advance the national HDI.

Then the HDI is contrasted against the village's resource security—measured as their Ecological Footprint compared to the biocapacity available to the village. Unfortunately, we did not have the resources to move beyond the mechanical prototype, using tablets for quicker and more effective data collection and use of the metric. But the participants from IDE-India, the villages, and the research team produced lively conversations using our diagrams, and village commitments ending up being painted on a public wall in the village.

The same project also tested this approach with Gramvikas. Believing that social justice is achieved not by talk but by action, Joe Madiath founded a powerful organization that uses inclusion as a way to build effective sanitation infrastructure, including toilets and a freshwater tap for every kitchen in low-income villages. These installations improve the health of the entire village and increase their agricultural and general economic productivity. Again, this is an example where higher resource security drives lasting well-being outcomes.[24]

CHAPTER 10

FOOTPRINT IN ARCHITECTURE AND CITY PLANNING

BedZED, Masdar City, and Peter Seidel

Bill Dunster and BedZED

British architect Bill Dunster has been working with Footprints for decades. He created his award-winning mixed-use housing project BedZED in Sutton, South London, in partnership with BioRegional. "Bed" refers to its location in Beddington, a community south of London, while "ZED" stands for "zero (fossil) energy development." Completed in 2002, BedZED is based on detailed Footprint calculations. The principles of ZEDliving focus not only on architectural issues, such as a building's insulation and heating.[1] More importantly, ZEDliving promotes the integration of living and working spaces with the aim of making commutes unnecessary, and addresses even the integration of the supply of food. At its core, the project is about a whole new style of living—as an opportunity, not an obligation. Not only in Europe but also in North America, in the Arab world, in Australia, and in South Africa, large sums of serious money are being invested in model developments that follow the principles of "One Planet Living."

"It was a while ago," Bill Dunster said, "I was an architect working for a large company and was specializing in low-energy buildings. At the time, we were building a lot of office spaces. One day it struck me that this work made absolutely no sense whatsoever as long as these office buildings were surrounded by enormous parking lots and we ignored the energy required to transport the people who worked in these offices, the energy that went into feeding them, and everything else. We simply did not ask the right questions."[2]

Bill Dunster's dream was to take into consideration the ecological impact of all of people's activities, from their choice of lunch to their commuting habits, their building's energy supply, right up to the question of how they spend their vacations. His idea was to look at the whole picture.

Bill Dunster got involved with the Ecological Footprint during the mid-1990s via Craig Simmons, a founder of Best Foot Forward and now the Chief Technology Officer of the Anthesis Group,[3] one of Europe's leading ecological consultancy groups. For Bill Dunster, the important frame of reference is not the global but the national—British—context. After all, the United Kingdom imports 70% of its food and is relatively densely populated. Against this background, Dunster and his team came up with a whole range of proposals for solutions.

The BedZED buildings were to be insulated so well that they would require only a minimum of energy, and heating would be gained from domestic biomass, especially waste wood. No energy was to be derived from food-producing areas. For the same reason, the team sought out a piece of land that had been previously developed and built on (brownfield land) rather than a site out in the country (greenfield land). BedZED was to offer employment for all its residents, that is, integrate work space with residential space, so that no energy or other resources would be needed for commuting. Also, residents were to be given private gardens, as is typical of British lifestyle. The entire development was to be built from local materials and be superbly connected to public transit. Finally, Peabody Trust,[4] a nonprofit organization in London with a history of social and ecological advocacy, agreed to take on the role of developer. Additional support came from the Bioregional group,[5] which, together with the World Wide Fund for Nature (WWF), wanted to showcase examples of "One Planet Living." Inspired by Footprint methodology, they asked, What would the world look like if it stayed within nature's budget?

All of this was unchartered territory. Preparations, in particular the planning of the countless details, were extremely time-

BedZED — The first larger-scale development ever with the explicit ambition to enable one-planet living conditions. The NGO Bioregional conceived and had the project completed by 2002. The designer was Bill Dunster.

Illustration: Phil Testemale

consuming. But in 2002, BedZED was finally completed. The complex with its three-story height comprises about 100 south-facing dwellings as well as north-facing workspaces; in addition, there are exhibition spaces and a kindergarten. Now that the project has been up for a while, we can confidently say that it has been a success: an affordable, attractive, and resource-saving pilot project with lots of class, color— even on the grass-covered roofs—and surprising shapes.

Immediately noticeable are colorful rooftop wind cowls. At first glance they look like chimneys, but they are actually part of the project's highly efficient ventilation system. They use the wind to draw warm stale air up from inside and to direct fresh air downwards. Because during the heating period the air inside is warmer than the air outside, heat exchangers inside the cowls transfer the heat from the air inside into the cooler air from outside so as not to lose the energy of the warmer air. Usually, this process requires external energy, typically electricity that drives the ventilation. But the trick of the heat exchangers in BedZED's cowls is their use of regenerative energy: they are driven by the wind.

A pedestrian route goes right through BedZED; in fact, the project has a "pedestrian first" policy. A pool of cars for shared usage also belongs to the complex. While privately owned cars are an option, it takes residents less time to use the bus or rail system, or to reach the car pool's parking lot. "It may not be a new thing, but it works," commented Bill Dunster. According to his philosophy, an ecological lifestyle must be attractive and ought to be promoted but not forced upon people. The architect distinguishes three groups among the current residents of BedZED: those who believe in an ecological lifestyle and whom he jokingly nicknames "eco-saints," those whose ecological commitment is limited to living in a passive house and who otherwise follow a more conventional lifestyle, and finally those for whom ecological responsibility basically means nothing.

When asked about the bottom line (namely the size of BedZED residents' Footprint), Bill Dunster said that it depends: "If someone is a vegetarian, works in the community, owns no car, and does not take plane rides to go on vacation, and if the renewable-energy systems work properly, then it is possible to have a One Planet lifestyle." The lifestyle of One Planet Living translates into achieving a Footprint of currently 1.6 global hectares—actually less if we want to keep some area for the sake of biodiversity. BedZED's architecture and way of construction, but above all its energy management and particular kind of interconnectedness between buildings and their environment, are all important parts of the whole. But an architect alone cannot pull off the entire change to One Planet Living.

Since 2002, Bill Dunster and his ZEDfactory have moved beyond their BedZED pilot project,[6] with Footprint and carbon-neutral thinking serving as their guide. Over the years, distinct ZED standards have been developed for settlements with different population densities. At the bottom end are rural settlements with up to 15 residential units per hectare. But ZED standards also exist for urban settlements and even for high-rise structures with up to 240 residential units per hectare. Typically, rural living means better availability of renewable energy and local food, while urban, higher-density living means shorter travel distances and more compact buildings. Different population densities create fundamentally different starting conditions and demand individual solutions.

Ecological construction, such as construction according to ZED standards, increases building costs, especially for early innovative and experimental projects, even if later operating costs are lower. "It's always cheaper to fob one's problems off to others instead of solving them in one's own backyard," Bill Dunster has said. One special advantage of BedZED is its clearly defined energy-consumption target in the hope of facilitating an attractive lifestyle with a Footprint that can be replicated globally. Much of BedZED, such as its rooftop wind cowls, is still at the experimental stage. But experiments are the only way we can learn. The resource efficiency of a small project with relatively few residential units, such as BedZED's, reduces Great Britain's Footprint only marginally. What is more important are the impetus and inspiration BedZED offers. The more honestly its results are analyzed, the more can be learned for future developments.

The challenges are considerable. If in 2050 we will indeed be nine to ten billion people, the average person will have a nature budget of only 1.2 global hectares, and if we were to reserve half of it for the sake of biodiversity, we'd be left with 0.6 global hectares per person. If everyone in the world were to have the same amount of biocapacity available, the British per person Footprint would have to shrink within a few decades considerably below the BedZED level.

Masdar City as Eco-City

At this point, a number of eco-cities are in the experimental stage. In Asia, hundreds of cities with over a million residents will be built over the next decades. So far, many of these projects exist only on the drawing board. But they all agree on the concept of One Planet Living, which the World Wide Fund for Nature (WWF) has promoted as well.[7] The design principles are, to list the most important ones: zero carbon, zero waste, sustainable transport, and local and sustainable materials.

While decisions regarding food and goods for daily use are generally up to consumers, issues of infrastructure, energy supply, and building density are essentially shaped by planners. Their decisions are, however, crucial for the level of a settlement's resource efficiency. This is why forward-thinking city design is key to sustainable life on this planet. The Footprint acts like a compass in those planning decisions. In 2016, ten One Planet Living projects were already either in the planning stage or fully operational, a list that includes projects from every continent.

Masdar City in the United Arab Emirates, too, is built inspired by One Planet Living principles.[8] The United Arab Emirates used to be known for palm-shaped artificial islands, underwater hotels, the tallest buildings in the world, and an enormously energy-intensive infrastructure. In such a context, Masdar City, a green city in the desert, is an experiment. The country's government knows full well what the world will have to face in terms of climate change and resource constraints. Thinking long-term, the oil sheikhs of the Emirates want to prepare for the time after oil production. Part of this long-term plan is Masdar City, a planned city originally proposed to span over two square miles and designed by British star architect Lord Norman Foster. The first sections have been built, and there's quite a bit that can be viewed. The city now houses academic institutions including Khalifa University of Science, Technology and Research.[9]

Masdar City has a high density and is car-free and pedestrian-friendly with narrow, shady streets. Solar energy cools the city. Buildings will not exceed five floors, and a person will never be farther than 200 meters away from access to public transport. Drinking water is produced in a solar-powered desalination plant. Green spaces within and cropland outside the city are irrigated with recycled water. Roofs are mostly covered with solar panels. The goal is for the city to produce within the city limits sufficient energy for its own consumption. A train connects Masdar City with Abu Dhabi, the capital of the United Arab Emirates, but within the city, residents travel on foot or use the small driverless cabs that run on tracks.

Masdar City in the United Arab Emirates was also inspired by One Planet Living principles—with the ambition to house 50,000 people. While currently only 10 percent is complete, the existing section showcases impressively what is possible, even in a harsh climate like the UAE.

Illustration: Phil Testemale

Realization of the project was somewhat slowed down by the financial crisis, and clearly not every problem has been solved in Masdar City. The vicinity of an airport is still seen as an advantage. But Masdar City is an investment in the right direction and responds to the certain knowledge that conventional city development creates structures that are so resource-intensive that they are bound to decrease in value. Abu Dhabi is now planning to build a nuclear power plant because its urban infrastructure is too energy-thirsty and the energy from the country's oil fields no longer suffices—a situation that shakes up even the technocrats in the United Arab Emirates' Governing Council. And Abu Dhabi is not alone; Dubai and other Emirates are planning coal-fired power plants.

Peter Seidel:[10]
A Sustainability Pioneer Looks Back

MATHIS: Peter, you had an incredible career as an architect and planner. What amazes me is your early recognition that our built infrastructure needs to become one-planet compatible. This was not obvious, even to your architectural teachers at the Illinois Institute of Technology. They were world-class designers and the founders of modernism. One was Mies van der Rohe,[11] the last director of the groundbreaking Bauhaus design school that sprouted modernist design around the world in the early 20th century. After escaping Nazi Germany and coming to the US, he built modernist masterpieces, including the aesthetically astonishing but ecologically uninformed campus of the Illinois Institute of Technology. How did you develop your ecological conviction?

PETER SEIDEL: When I studied in Chicago in the 1950s at this remarkable school, three teachers came from Bauhaus. Beyond the architect Mies van der Rohe, I studied with urban planner Ludwig Hilbersheimer, and photographer and visual artist Walter Peterhans. It was a tremendously stimulating atmosphere!

I remained in Chicago, starting my career in the late 1950s as an architectural designer. Surprisingly, there had been architects in Chicago who had built solar houses in the 1930s, and there were a number of architects who were interested in it, but society's lack of curiosity did not allow these ideas to flourish.

MATHIS: What turned you into an ecological architect?

PETER: I was part of the architectural team for the Air Force Academy, which were extravagant buildings in the foothills in Colorado, right on the edge of the mountains. I was working on the housing part, and I felt the whole project didn't fit at all. These buildings were glass boxes designed in Chicago by people who were not much involved with nature and mountains. We would build air-conditioned industrial buildings in cool, colorful Colorado! It was such an absurd contrast.

At that time, someone handed me *The Next Hundred Years: A Discussion Prepared for Leaders of American Industry* published in 1957 by Harrison Brown, James Bonner, and John Weir.[12] I also devoured

The eco-settlement Vauban in Freiburg, Germany, is one of the most recognized large-scale development in Germany embracing sustainability.

Illustration: Phil Testemale

The Challenge of Man's Future by Harrison Brown,[13] emphasizing two main topics that still stick in my mind to this day. One was population growth, an important driver few seem to pay serious attention to. It also talked about resource constraints and how they could lead us into a lot of troubles in the future if not attended. That book totally changed me. I saw that everything I was doing was wrong. I was working in Colorado on buildings that were energy hogs and utterly detached from their context. After all, this was the Air Force Academy for young cadets—a place that trained future leaders. This and the glass offices we built were such misfits! I revolted.

When I saw those buildings we were working on, and the damage they were doing to the planet, I felt that I was on the wrong track and that I ought to switch. I had to change; there was no question if I had any personal integrity. So I decided to become an environmental architect.

MATHIS: What did your friends and family think?

PETER: My family is very conservative. They love nature. But thinking or doing something radically different was outside of the way my family thought. Therefore, I decided to start teaching. I was in Western Virginia, and I got very interested in city planning with an environmental bent. One source of inspiration was my teacher Ludwig Hilbersheimer. One of his great ideas was to put all the cars at the bottom of Lake Michigan. I did spend quite a bit of time thinking. I got into the business of designing environmentally sound new communities.

Then I transferred to the University of Michigan to teach architectural design. I was approaching each project *my way*. I was working quietly with my students. Some of them appreciated, but I also know that some didn't like it. There was some revolt from the students against my approach to learning from projects in the classroom, and faculty backing them. During that time, besides my ideas for environmentally sound communities, I became fascinated by solar energy and waste recycling. Thankfully, I wasn't alone. I moved to Cincinnati, Ohio, in early 1969 where I was hired to design a new town for a builders group.[14] Unfortunately, that project fell through. Nevertheless, there was a growing interest in the environment and in developing solar homes. Jimmy Carter had put solar panels on the White House (which Ronald Reagan ripped off as soon as he was elected President). But the demand for ecological houses was also limited, even though by then more architects were able to provide solar power homes.

In this context, it dawned on me: We know what's happening to the planet; we know what our responsibility is, and we know what we can do about it. I'm a visual not a verbal person. But as I kept pondering about these things, it dawned on me "My gosh, I've got to say it." Which led to my first book in the mid-1990s, *Invisible Walls: Why We Ignore the Damage We Inflict on the Planet—and Ourselves*.[15]

MATHIS: This was such a fabulous book; I believe the first ever to so clearly lay out the psychological and cultural barriers we are up against in the transformation to sustainability. And that's how I got to know about you, Peter. Now, you are 92 years old. What is your conclusion?

PETER: I have no regrets. The choices I made were the right thing to do, and I knew it. You have a goal, you set something up, and then you plot ahead and do it. That's how it unfolded.

CHAPTER 11

CHINA

A New Model of Development?

The Chinese leadership has come to realize that economic growth alone will not lead China into a prosperous, stable future. The ecological damage caused by China's rapid economic development has become all too visible. Immediate pollution alone has started to threaten economic and social development itself. The larger resource dimension is even more daunting. This huge country's Footprint has grown to 3.8 Chinas, up from 2.2 Chinas in 2000. For several years now, China's government has a made serious attempt to address the country's challenges. For instance, their 13th Five-Year Plan (2016–2020) has on average five references per page to nature, ecology, environment, energy, land, water, or resources.[1]

The Chinese national government proposes two key strategies in response, both linked to what otherwise would be called a sustainable development agenda. The "One Belt, One Road" initiative connects China more deeply with the rest of the world; "Ecological Civilization" advances harmony between people and nature. While the first one may be more focused on increasing China's resource security through building physical access, the latter one recognizes sustainable development as a necessary ingredient for China's own economic success. Yet, the difficulties of that path cannot be overestimated. China's development has already been a balancing act for some time: industrialization, urbanization, economic restructuring toward a market economy, and an opening-up to the world market—and all of this has happened with a speed hardly imaginable in any other part of the world.

A detailed Footprint study, which Global Footprint Network produced in collaboration with a provincial government, applied ecological accounting methods to the conditions in China.[2]

Throughout the past fifty years, China has experienced massive material expansion. Its population has doubled since 1961. However, over the same period, its *per capita* Footprint has nearly quadrupled in size. As a result, China's demand for biocapacity has grown eightfold. At the beginning of the 1960s, China's per capita Footprint placed it in 114th position on the list of nations. Today, it is in the 66th position for per person Footprint.

China's total country Footprint is quite a different matter. China surpassed the United States' country Footprint in 2003, and sports now a country Ecological Footprint twice as large as the US, and three times larger than India. Also, its biocapacity deficit—not its *per capita* one, but that of the country as a whole—is now the biggest in the world. It is 2½ times bigger than the US biocapacity deficit, the second largest in the world. China's deficit is also seven times larger than India's (in third place), and fourteen times larger than that of Japan and Germany, with the fourth- and fifth-largest biocapacity deficits, respectively.[3]

With current growth rates, China's gross domestic product doubles every seven to ten years. The chart on page 105 shows the trends of China's Footprint and biocapacity since 1961.

When Deng Xiaoping started his 1978 era of reform and opening up, he kept the communist state-run agriculture and industries but built around them de facto private industry. Since then, the relative size of the state-run economy has significantly shrunk. The concomitant labor market shifts are gigantic. Every year state-run industries lay off millions of people, while new entrants crowd the labor market. New jobs can only come from the private sector—an enormous social experiment without a safety net. Traditional family structures carry most of the burden and offer a minimum of social redistribution.

The Chinese catch-up process is an unprecedented tour de force. The Chinese are compressing industrialization and urbanization—processes which took about two centuries in Europe—into a few short decades. At the same time, China's transformation into a market economy is going full steam ahead. The speed of development, especially of the booming regions in the country's south, leaves many Western observers dizzy. Shanghai is well on the way to overtaking New York as the world's most glitzy metropolis.

But TV images of rising China should not let us forget that the still about ½ of China's residents live in the countryside. Government intervention has prevented a massive rural exodus with all its consequences, such as the risk of mass impoverishment in the cities. Yet there may be no other country where urban-rural polarization is growing as fast as in China. But even within cities, the income differential is spreading. The focus on ecological civilization may also be caused by the Chinese government's concerns that environmental degradation and conflicts over resource constraints may trigger social unrest. So far, its answer to the problem has always been the same: economic growth.

With all of China's efforts, direct pollution impacts are still severe. Researchers from the Chinese University of Hong Kong estimated that air pollution in China leads to 1.1 million premature deaths in the country each year and is destroying 20 million tonnes of rice, wheat, maize, and soybeans.[4] Many Chinese cities suffer from considerable air pollution. Respiratory diseases, including lung cancer, are topping the statistics. The main cause of air pollution is the widespread use of domestic coal. One third of the country suffers from acid rain, the result of sulfur dioxide and nitrogen oxides emissions. In addition, northern China is experiencing water scarcity.[5] It is made worse by the dramatic pollution of Chinese rivers. Clean drinking water is a rare commodity in China. Climate change is expected to lead to a complete melting of the Tibetan glaciers and to stronger typhoons. China is hence both a perpetrator and a victim of climate change.

The gap between the consumption standards of different Chinese populations is enormous and keeps growing, but if we calculate the mean score of their consumption, we arrive at the previously mentioned Footprint of 3.6 global hectares per person. As such, the demands of the average Chinese are way lower than those of a European or Australian. But if people in China or across the Asia-Pacific region eat just a little more meat, have the occasional beer, and drive a car instead of riding a bicycle, it will shift the resource balance of the whole planet. It will also affect China's own resource risk.

However, China also has great natural wealth. The biocapacity of Chinese cropland is the second highest of all nations. That of Chinese grazing lands surpasses that of all OECD countries' grazing lands taken together. All these numbers make clear how important today's carbon Footprint is for China: it makes up over 69 percent of China's demand for biocapacity. China, like many other countries, cannot absorb so much CO_2 with its own biocapacity; this means that global ecosystems must deal with those emissions. Compared to global trends, the changes indicated by the Chinese Footprint and biocapacity figures are extraordinarily rapid. As already mentioned, China has lived with a biocapacity deficit since the early 1970s but also with its domestic, albeit limited, means of compensating for this deficit through resource imports.

Also in the early 1970s, ecological constraints became apparent. It was therefore no accident that a few years later the Chinese leadership introduced its one-child policy, a drastic policy hardly imaginable in other parts of the world. A far more slowly expanding population size significantly reduced China's risk of suffering famines, as it repeatedly had suffered in the past. Chinese *per capita* biocapacity is about 17% lower than that of the Netherlands, and about 40% lower than that of Germany. But high-income countries are better able to afford imports than are low-income ones.

In the Pearl River Delta region, Hong Kong, Shenzhen, Guangzhou, Zhuhai, and Macau have merged into one 40-million megacity. Not too long ago, herons lived in the delta's swamps and fishing villages lined its banks. Today, within a radius of 60 miles, this region includes five international airports and 41 ports. They are the shopping paradise of IKEA, Walmart, and Apple. Every major electronics producer is located there.

China's trade statistics show an abundantly clear pattern: raw materials are imported, and manufactured products are exported. For example, China buys wool in Australia and New Zealand, produces fabric and garments domestically, and exports them, mostly to the United States and Japan. Taken together, Chinese imports in 2016 represented about 880 million global hectares of biocapacity. On the opposite side of the ledger were exports worth 590 million. The difference, consequently, is 290 million global hectares. This, too, explains China's considerable biocapacity deficit.

All this shows the degree to which China's development depends on a resource flow entering the country. Over centuries, feudal China had lived in self-imposed isolation with a by-and-large self-sufficient economy. But today's China needs stable and secure trade relations as much as we need the air we breathe, and it will need them even more in future. From the United States, China imports predominantly grains, wood, and fiber, not least for paper production. Australia in turn is one of the main customers of Chinese paper. The resource flow both in and out of China is impressive.

Footprint studies already have some history in China. The first one was done in 1999, and several dozen exist today, as do countless academic publications put out by Chinese universities. The results of this research are reflected in government decisions. In essence, China's resource security is becoming more uncertain because of climate change, price volatility, the emergence of new energy technologies and eventual resource constraints. China's enormous resource dependence is a particularly critical factor.

As the future of fossil fuel is clouded, reducing its use might significantly increase the demand on other resources, which are limited as well. China's large dependence on fossil fuel and biological resources for key industries and for developing and powering its cities—including gasoline for transportation and coal for electricity, food, water, and fibers—is a major vulnerability. It is a challenge that China has the knowledge and economic power to overcome. The question is: Will China focus on prioritizing "Ecological Civilization" that emphasizes harmony with nature over "One Belt, One Road" strategies that bank on accessing more resources from abroad?

Particularly in the face of the rapid speed of infrastructure development in China, focus on the impact of long-lasting assets is critical. Because of the long lifespan of built infrastructure, ensuring new infrastructure investments are suited for a resource-constrained future is imperative. This is where Ecological Footprint assessments become valuable, as they help identify options that increase resource security while also being economically and politically desirable.

On some level, China recognizes already that, as population growth and the rise of standards of living around the world continue intensifying human demand on natural resources, an economy in harmony with nature is the only path to future resilience. Such an economy will become any country's most valuable asset. Given the size of China, can it adjust to these new realities fast enough?

CHAPTER 12

AFRICA

Protecting One's Resources

Wood, beef, oil, diamonds—in an increasingly resource-constrained world, many industrialized or newly industrializing countries are turning to Africa to satisfy their desires. Eying the Footprint map of the African continent, we realize that Africa's long history as a biocapacity creditor has recently come to an end. Many African states have already slipped into a biocapacity deficit. With a relatively constant per person Footprint but a still rapidly growing population, the ecological situation is becoming critical.

From the Footprint point of view, the continent's prospects are clear: regions and countries must seriously protect and wisely manage their ecosystems to secure their own existence.[1] For it will become ever more difficult to buy additional biocapacity, especially for low-income populations.

In 2016, 16% of the global population lived in Africa. But Africans contribute a mere 5% to the global Footprint. If, in that year, everyone in the world had consumed at the same level as Africans, humanity would not be using 1.75 but just half of that: 0.85 Earths. Why would Africans have to worry about overshoot?

From a Footprint perspective, Africa is in a peculiar situation. Its people's *per capita* Footprint has stayed relatively unchanged for years and sits currently at 1.4 global hectares, a notably lower level than

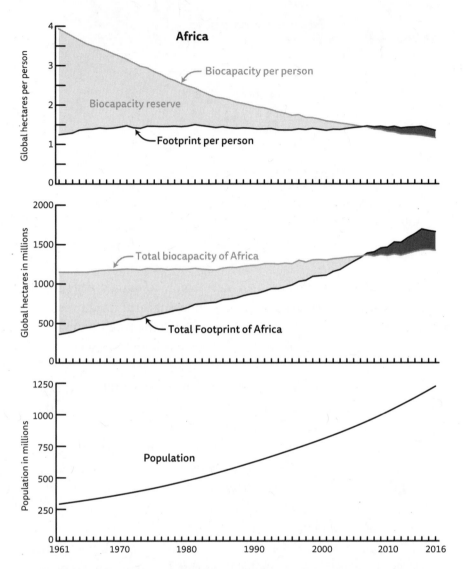

Figure 12.1. Africa's Footprint and biocapacity, per person and for the entire continent. Bottom, Africa's population over time. Credit: National Footprint and Biocapacity Accounts.

that of many other regions of the world. Per person, Africans claim about ¾ the global average of biocapacity. But for anyone associating Africa's nature with wide savannahs and lush rain forests, it must be sobering to realize that the continent's biocapacity of 1.2 global hectares per person is way below the global average of 1.63 global hectares per person in 2016.

The crucial reason for this meager *per capita* supply of biocapacity has been the extraordinarily rapid population growth over the past decades. From 1961 to 2019, the number of Africans quadrupled from 300 million to more than 1.3 billion.[2] During that same period, the continent's per person Footprint shrank from 1.35 global hectares per person to 1.14. Global Footprint Network estimates indicate that the continent is now approaching a biocapacity deficit. This must be for the first time in its history.[3]

Conservative projections by the United Nations assume that the population of Africa might reach 2.5 billion people by mid-century. According to their median projection, more recent UN prognoses conclude that there might be 4.4 billion people living in Africa by the end of the century.[4] Is that even physically possible? Already many African states are biocapacity debtors. Given global resource constraints, resources can be expected to become more costly and in fact unaffordable for countries with low *per capita* incomes. This underscores the need to focus on resource security in order to secure lasting development success. After all, rapid population growth eats up opportunities first and foremost for local populations. Ignoring these trends undermines the possibilities and dignity of the African generations to come.

Take Niger, for example, one of the lowest-income countries in the world. It is hot and dry; in fact, by far the largest part of Niger is desert. In the Sahara regions, agriculture is only possible in oases created by artificial irrigation. Where the Niger river runs through the country's southwest, a short rainy season exists, but precipitation is inconsistent. The main crops of Niger are different types of millet as well as beans and peanuts. The overwhelming majority of Nigeriens live self-sufficient lives in rural environments, and with a sharply growing population, even a minimal buffer of available biocapacity is easily lost. For more than 25 years, Niger's Footprint has been limited by the country's biocapacity. In 2004 and 2005, droughts and locust infestations led to considerable crop losses. Another locust invasion hit in 2012. In addition, the population grows at about 3% per year, meaning a doubling of less than 25 years. In fact, since its independence in 1960, the population has grown more than sixfold.

As supply (biocapacity) no longer meets the demand (Footprint), there are financial constraints, and food security is a permanent crisis—accentuated by biocapacity fluctuations due to weather or pest. All these trends put Niger's development path enormously at risk, particularly given that current trends suggest that its population is expected to again quadruple before the end of this century—something unimaginable from biocapacity perspective.

Given the variety and differences among regions and countries on the African continent, Footprint analysis of the African continent as a whole cannot do justice to the situation. With the help of the United Nations data sets, we can trace the resource situation of most of Africa's 54 countries individually, all the way back to 1961.

The insights from such resource analysis profoundly contradict mainstream development approaches and policies. By distinguishing countries' economic development patterns as a function of their natural capital performance, the Footprint's view is filling a blindspot of current competitiveness and sustainability research. Footprint accounting demonstrates that resource security, and not economic expansion, is the critical enabler of lasting development, including poverty eradication. Given, that, at least according to the World Bank and IMF data, African countries' income per person is still rising (although slowly), this biophysical view also indicates that conventional measures, such as income, are unable to detect the fundamental resource trap.

We offer the following sketches of our Footprint developments in a few African countries to make visible these otherwise hidden relationships.

Egypt

Egypt's population is approaching 100 million people. This is about four times more than its population in 1960. With intensification of agriculture, including irrigation and fertilizers, Egypt's apparent

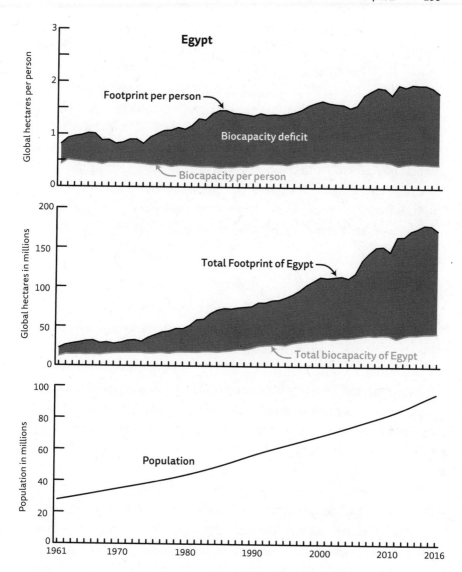

Figure 12.2. Egypt's Footprint and biocapacity, per person and of the country as a whole. Bottom, Egypt's population over time. Credit: National Footprint and Biocapacity Accounts.

biocapacity per person has remained nearly constant, just 0.45 global hectares per person. Irrigation heavily depends on the Nile River which provides 97% of the freshwater to Egypt, compared to 3% coming from domestic rainfall. Of course, all freshwater comes from rainfall, just that 97% does not fall on Egypt's territory, which underlines a challenge to Egypt's resource security. Still, the 0.45 global hectares per person of biocapacity meet an ever-increasing Footprint which has grown from 0.8 (in 1961) to 1.8 (in 2016) global hectares per resident. Egypt was already in significant biocapacity deficit in 1961. But since, Egypt has expanded from using two Egypts to using four—or many more if we include the areas needed to catch Egypt's freshwater.

Algeria

From 1961 to 2014, Algeria's per capita Footprint more than doubled from 0.73 to 2.4 global hectares. Its population, too, quadrupled during that period. In 2017, the country had 41 million people. The country's biocapacity per person, however, showed the opposite trend: it sank from 1.44 global hectares in 1961 to 0.5 global hectares in 2016. In this process, Algeria became increasingly dependent on imported biocapacity. Eventually, the country's biocapacity matched only ⅕ of its Footprint.

Kenya

By 2018, Kenya had surpassed the 50 million people population mark, six times more than Kenya housed in 1960. Its biocapacity per person sank from 1.8 global hectares in 1961 to 0.5 global hectares in 2016. The Ecological Footprint per person shrank in parallel: from 1.7 global hectares in 1961 to 1 global hectare per person. Even though Kenya's Ecological Footprint per person shrank, their biocapacity deficit increased from a slight reserve to now using twice the biocapacity that Kenya has available domestically. Kenya exhibits being stuck in what Global Footprint Network labels an *ecological poverty trap*. Limited biocapacity and limited purchasing power shrink the country's low

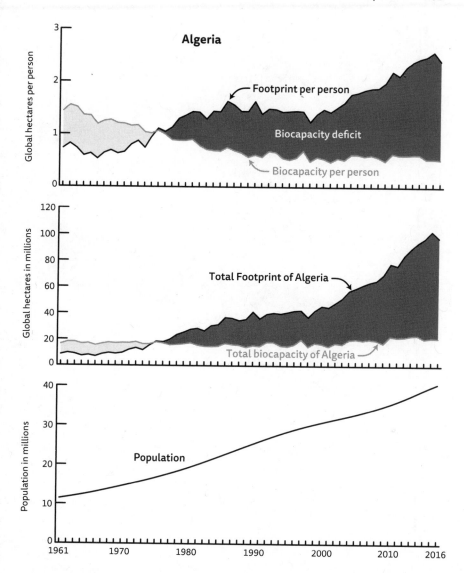

Figure 12.3. Algeria's Footprint and biocapacity, per person and of the country as a whole. Bottom, Algeria's population over time. Credit: National Footprint and Biocapacity Accounts 2019 edition.

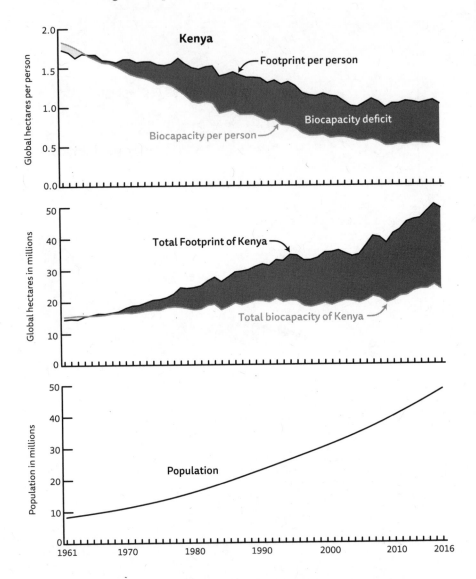

Figure 12.4. Kenya's Footprint and biocapacity, per person and of the country as a whole. Bottom, Kenya's population over time. Credit: National Footprint and Biocapacity Accounts 2019 edition.

per capita Footprint even lower, with few opportunities, given current policies, to reverse the trend.

Mali

At the time of writing, Mali was just about to enter a biocapacity deficit situation, with a Footprint and biocapacity of 1.5 global hectares per person. From 1960 to 2018, its population nearly quadrupled to 19 million people. Over that same period, its biocapacity per person dropped by 55% from 3.3 to 1.5 global hectares. In 1961, Mali's biocapacity had been almost three times as big as the country's Footprint.

Mozambique

For a while, Mozambique's biocapacity was nearly eight times bigger than its biocapacity in 1961. Today, it is still double. At a low Ecological Footprint level of just 0.8 global hectares per person, Mozambique's *per capita* Footprint remained relatively unchanged from 1961 to 2016. During that same period, the country's population nearly quadrupled. As a result, its *per capita* biocapacity dropped from more than 6.6 global hectares in 1961 to nearly 1.8 global hectares in 2016.

South Africa

South Africa had a Footprint peak in 1974 and 2008 with 4 global hectares per person and 3.9, respectively. Now it stands at 3.15 global hectares per resident—a decline driven more by economic challenges than astute sustainability policies. South Africa's biocapacity fell from 3 global hectares per person to 0.96, largely because the population more than doubled in this period. As a result, South Africa now requires over three South Africas.

Tanzania

In 1961, Tanzania's biocapacity exceeded its Footprint by a factor of two. Today, it is running a slight biocapacity deficit. Recent trends have been quite staggering: Tanzania doubled its cropland productivity in the 10 years after 2008. This recent upswing reversed Tanzania's

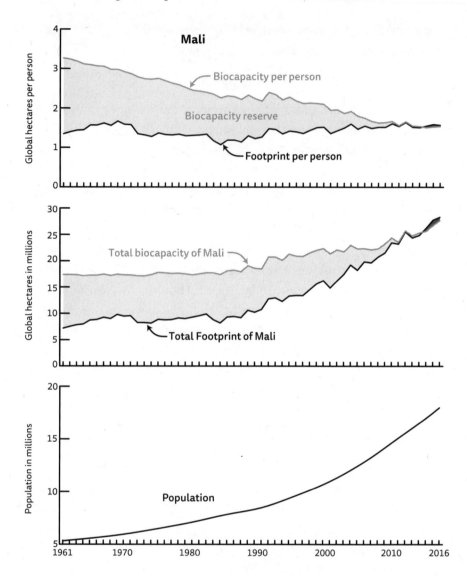

Figure 12.5. Mali's Footprint and biocapacity, per person and of the country as a whole. Bottom, Mali's population over time. Credit: National Footprint and Biocapacity Accounts 2019 edition.

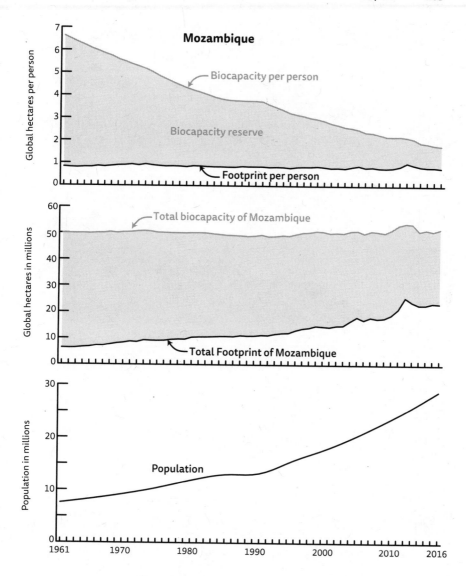

Figure 12.6. Mozambique's Footprint and biocapacity, per person and of the country as a whole. Bottom, Mozambique's population over time. Credit: National Footprint and Biocapacity Accounts 2019 edition.

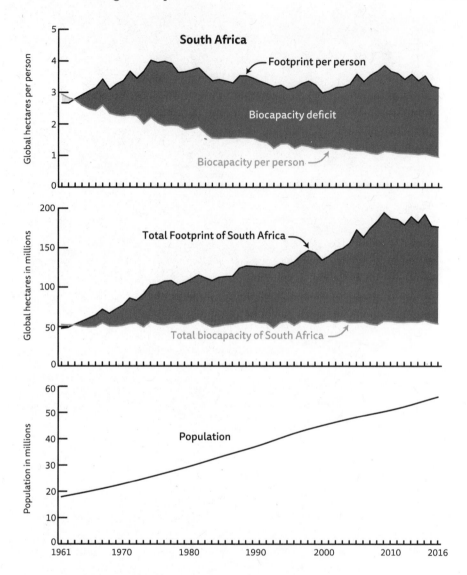

Figure 12.7. South Africa's Footprint and biocapacity, per person and of the country as a whole. Bottom, South Africa's population over time. Credit: National Footprint and Biocapacity Accounts 2019 edition.

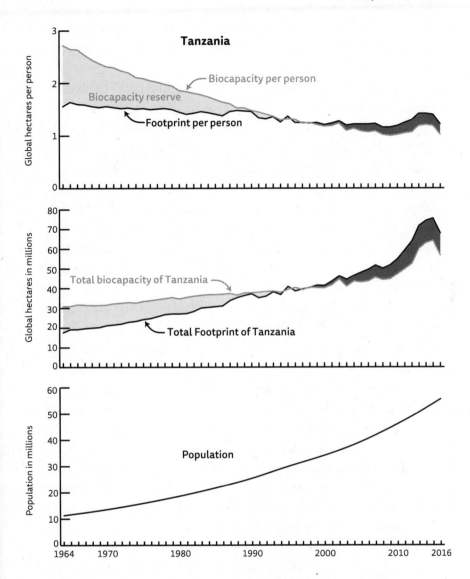

Figure 12.8. Tanzania's Footprint and biocapacity, per person and of the country as a whole. Bottom, Tanzania's population over time. Credit: National Footprint and Biocapacity Accounts 2019 edition.

per capita Footprint, for a few years, but in 2016 it was down again to 1.2 global hectares per person. Biocapacity declined from 2.7 global hectares per resident down to 1.

From 1961 to 2016, the biocapacity of all of Africa increased by about 34%; however, the continent's population more than quadrupled over that same period. Many African countries do not have enough biocapacity to provide for their populations, nor the financial resources to buy it from elsewhere. Other countries like Botswana or Gabon are better positioned. But the fact remains that millions of Africans depend on their local, renewable resources, be it fuel for cooking or foodstuff such as fish, tubers, or grains. In many countries on the African continent, a large percentage of their labor force works in agriculture. Also, exporting biocapacity in the form of agricultural products is a crucial source of income for many African countries. Africa's future will hence be significantly shaped by how well, or how poorly, the continent manages its resource security. This is again a place where Footprint accounting can assist: it makes obvious the relationship between human demand and what ecosystems can renew.

Regionalizing data and the monitoring made possible by such data would be very helpful to African countries. It could, for example, help to secure adequate supplies for the continent's growing cities, megacities, and its informal settlements.

The widespread trend of people leaving rural communities for big cities continues unabated in Africa. However, city residents also need grain, wood for fuel, and fish. And water and energy. If such resources are unavailable, human tragedies can quickly develop. With no other options, populations will help themselves to the resources of their region, in the process of which, ecosystems tend to get overexploited. But the fact is that also the rest of the world does not have a biocapacity reserve. On the current path, overexploitation becomes inevitable. Preventing overexploitation and shifting humanity's course is becoming a foundational necessity if we want a bright future. How much do we have? How much do we consume? Here? Elsewhere? As humanity?

Pressure on African biocapacity does not come only from Africans themselves. High-tech international fishing fleets, for example, have for years been scouring traditional fishing grounds along the west and east African coastlines, often to the point of fish stocks' collapse. Meanwhile, domestic fishers return with empty boats. Many of the once lively and abundantly stocked fishing grounds along the African coasts have been left desolate.[5]

Another very real danger for many African countries comes from illegal logging. At least in the early 2000s, due to mismanagement of Tanzania's forest industry, China, for example, is thought to import ten times more timber from Tanzania than Tanzania's official statistics show.[6] The result is lost revenue for Tanzania, as well as pressure on its forests.[7] Illegal logging can destroy entire forests and subsequently lead to erosion, floods, and changes in the local climate—none of them good conditions for the local population.

Also, private as well as public investors from high-income countries are using land in Africa to produce food.[8] China, for example, has leased 2.8 million hectares in the Democratic Republic of Congo to grow palm oil. Egypt secured for itself hundreds of thousands of hectares in Sudan to grow wheat. Libya made large-scale investments in Mali, also to grow wheat. Local farmers are driven from the fertile Niger delta—typically to places with less water and poorer soils. Given political unrest in Libya and Mali, the respective contracts probably did not hold.[9]

All these massive deals carry a huge risk of pushing large numbers of people into poverty and starvation. Large investors bag the harvest and leave little behind, and so the pressures increase. Investors' hopes for stable future yields remain unfulfilled, and even less hope exists for the local population.

The future of African countries lies in the preservation of their own ecosystems, of their forests, fields, and fishing grounds, and not so much in their tourism industries (such as Kenya's). When the cost of air travel goes up, investments in five-star hotels are lost. In this regard, as always, Footprint accounting provides twofold support:

it takes a realistic inventory, and it gives cause for hope by showing paths forward.

Approaches as espoused by organizations like Blue Ventures,[10] who combine marine conservation with community development and the empowerment of women, or CAMFED—Campaign for Female Education,[11] who have enabled over three million girls in southern Africa to go to school (leading to a measurable reduction in family size of the graduates), are exemplifying what is needed: combining resource security with advancing opportunities for better lives for all.

CHAPTER 13

FOOTPRINT

A Conversation

Impossible Trends

BERT: What is the Ecological Footprint all about?

MATHIS: When I was born, humanity used ¾ of the biocapacity of our planet. When my son was born in 2001, we were already using 1.38 times what planet Earth renews. Today he is 18, and we require 1.75 times the Earth's regenerative capacity. Once he reaches the age I am now—and if the United Nations' conservative projections for population growth and energy demand come true—humanity would claim three times the amount our planet can regenerate.[1]

BERT: Sounds crazy...

MATHIS: And it's in all likelihood physically impossible. But these are the trends and the hopes. Hardly any economically focused institution asks serious questions about those trajectories. More effort still goes into devising ways to expand our economies. Even most climate scenarios assume massive economic expansion, while magically reducing emissions. Countries, cities, and companies apparently cannot imagine any alternative

Illustration: Phil Testemale

to growth and develop plans to take them in exactly the direction of even more massive ecological overshoot. Maybe a more productive way to address this conundrum is by asking: Will our investments keep their value in an ecologically severely constrained world? Some will, but others won't. Blindness in the face of these trends will only lead to nasty economic surprises. But it doesn't have to be that way if we prepare ourselves. We could create quite a pleasant future for ourselves if we really want to face up to physical realities.

BERT: The global population of about 7.7 billion today is projected to increase to between 9 and 10 billion by mid-century.[2] Newly industrializing countries such as China, India, and Brazil are catching up economically. But Footprint calculations show that today's crew of spaceship Earth is already consuming more than the planet regenerates. That doesn't exactly sound encouraging.

To Thrive Within Nature's Budget

MATHIS: I find sustainability conversations sprinkled with mixtures of hope and despair rather unproductive. If we beg for hope, we imply that our situation is hopeless. We thereby move the discussion into a religious discourse. As an engineer, I like to look at things practically. Engineers are not asked about hope and despair when they calculate the strength of a bridge. Calculating the strength of biocapacity is similar. That's why Global Footprint Network pursues a pragmatic approach. I don't believe humanity wants to commit suicide. People want to live, and live well. And they don't want to go bankrupt, either ecologically or financially. I am convinced that it is possible to thrive within nature's budget. Technically, it is possible; there are plenty of solutions. But we are a long way away from following a truly sustainable path. If we wanted to live sustainably, we'd have to fundamentally change our economic models—for example, by shifting power from financial to natural capital, by increasing taxes on resource consumption and waste production, or by promoting innovations that make us one-planet compatible. The bottom line is clear: business as usual is becoming not just an expensive, but a brutally destructive proposition.

BERT: But how can we create change?

We Need the Right Kind of Information

MATHIS: Denying the problem and putting our heads in the sand makes everything more difficult and takes us closer to the abyss. The right response is to observe with realism and act with pragmatism. That is why I have put much focus on taking a good look at our situation, and make efforts to describe it as clearly as I can. To me, this is the reason why ecological accounting matters. It is robust and gives us foresight. It doesn't make sense to wait blindly, and only change course after bankruptcy. We have the ability to anticipate. And we also know that our physical systems are slow at turning around, much like large supertankers. We need realistic information early on. Using the balance sheets of our ecological accounts, we have the power to determine now which ecological expenses we need to avoid and which revenues we need to increase if we want to avert ecological bankruptcy.

BERT: The Footprint has concluded that, in purely mathematical terms, 1.6 global hectares of biocapacity is presently available for every person on the planet. To people who live in high-income countries, such as Canada, the US, or Switzerland, this number sounds like an unreasonable demand. Currently, their Footprints are claiming way more.

Everyday Madness

MATHIS: I am witnessing two diametrically opposed experiences of the world. On the one hand—and this is true for me as well—people in affluent areas such as San Francisco, Hong Kong, or Hamburg lead lives with incredible material blessings: fancy food, heated swimming pools, air travel all over the world. Ergo, *We've never had it better*.

At the same time, we know we would need three to six planets if everyone in the world lived the way I and all those people do. That upper limit applies equally to me. With all my professional traveling, I have been in, if not exceeded, the six Earths zone—in spite of bicycling to work. I also realize that a large majority of people want to live the kind of life most people in San Francisco and Hamburg have access to. Hundreds of millions in China, India, and Brazil are ready to emulate those lives. The tide can hardly be turned. With these swooping tides of growing resource demand, we are inevitably devouring our future. This other view says, *We're doomed.*

Amazingly, these two views do not represent two separate camps, such as Republicans versus Democrats. Instead, many influential people in the industrialized parts of the world hold both views simultaneously. Including me. We live in confusion, with lack of integrity, without a vision we truly believe, all leading to denial. Only autistic kids like Greta Thunberg[3] can see through and verbalize it with clarity.

BERT: Put differently, the world is going to hell, but somehow my family and me, we'll manage. Al Gore spoke of an "inconvenient truth." But isn't that "inconvenient truth" somehow part of the Footprint as well?

We Are Making Ourselves Go Under, All of Us Together and Everywhere

MATHIS: Sure there is some inconvenience. But ignoring this becomes even more inconvenient. I am afraid that by framing our challenge as a carbon emissions problem, the debate focuses too much on this "tragedy of the commons." Most people recognize that it would be far better for humanity to have fewer carbon emissions, but their own emissions allow them cheap energy access for their personal comforts and thrills. This turns reductions into a service to humanity with few direct benefits to those who reduce their emissions. In contrast, when we all recognize that our challenge is rooted in the constraints of our planet's biological capacity, the dynamics can shift. It becomes obvious that it is in the self-interest of countries and cities to address the squeeze. Anyone unprepared will be exposed to being resource insecure. Inaction is self-defeating and becomes a security risk.

BERT: So you are flipping the story from inconvenience to self-interest?

MATHIS: Yes, exactly. We cities, countries, or companies are misjudging our self-interest. Yes, our current climate debate does not make obvious that it's in our direct interest to prepare for a future we can reasonably well predict. If you are not prepared, YOU are not prepared. Why are you satisfied with an ill-prepared boat for the upcoming storm? I'd be quite surprised if you and I won't see a good part of this storm in our lifetime. Good accounting helps us to see those possibilities more clearly. It shows that sustainability is becoming an economic necessity.

BERT: What does self-interest have to do with accounting?

MATHIS: The accounting argument may be easier to understand in the context of money. Operating a business or our lives becomes much more difficult if we don't establish realistic budgets. Without the clarity accounts give us, we are taken by surprise, and our finances can turn into a mess. The same applies to our resource situation. The ecological bank statements of countries—our National Footprint and Biocapacity Accounts stretching over the past fifty years—reveal that on average demand for resources has been going up, while the world's *per capita* biocapacity has been falling. These trends have been the same for almost every country. In fact, it's been such a universal trend that we assume this situation to be normal. But it's not. We are running toward a brick wall, all of us together. There are not even a handful of places where the trends go the other direction, far too few to save us. This is eerily similar to the story of the *Titanic*: there are far too few lifeboats while we steam confidently ahead into the iceberg zone. We are putting ourselves at risk, simultaneously. Not because it is inevitable, but because we refuse collectively to recognize that we are physical beings on a physically limited planet.

I am more than 56 years old. My entire privileged life has always been pampered by the same comforts: ever since I was born, the houses I lived in had electricity, running water, heating, a fridge. Yet this "always" is an illusion; historically speaking, 56 years is a short period. So, the current situation is by no means normal. That's why we say it might not be a bad idea to look from time to time at nature's bank statement. Are we in the black or in the red? In what direction are we heading?

BERT: If the Footprint identifies our ecological constraints, doesn't it start to look as if it were telling people how they should live? For example, that they shouldn't fly, or at least not so often?

MATHIS: Footprint accounting helps people to better understand our ecological context and what the implications might be for their countries and cities. Going over limits has consequences, as we well accept in other parts of our lives. How much money do we have? How big a property or farm can we afford? How much time do we have in a day? How long can I stay in this hotel? How much can I eat? How long can I stay underwater? Maybe we accept those limits because we experience them more directly. We live with them day in and day out and don't object. Some limits are even liberating. On a property without fences, children are more fearful and play closer to the house, but on a fenced-in property, they play all over the yard. Limits give safety, clarity, and structure. Ecological limits are just a description of reality, and not something people have invented and forced upon others.

Describe, Don't Moralize

BERT: When does communication about the Ecological Footprint model work well, and when does it not?

MATHIS: That's the biggest challenge, I think: how do you best communicate observations that challenge humanity's self-image? One big mistake of the environmental movement has been its moralistic or even moralizing stance. You may be right, but you don't make friends.

BERT: ...when you wag your finger.

MATHIS: How inviting is it to say: "The world would be better off without you, and since you already exist, consume half!" That's what many hear, and few are deeply inspired by this. Many may just quietly think: If this is so, I'd better look away and make sure I do not lose my privileges...To truly invite, we need to offer a different proposition and communicate that we want everyone to be part of this adventure. We need to be honest that we are facing a big problem, but at the same time, people need to hear that they are wanted and needed, that their creativity and participation is truly cherished.

BERT: How do you invite?

MATHIS: This is a key question to which I do not yet have a fully satisfying answer. But I know that it has to be at least as enthusiastic and genuine as inviting people to my birthday party. I really want people to come and come with joy. At Global Footprint Network, we try to invite by "building networks of robust inquiry," by doing research together, and by cooperatively looking for constructive opportunities. In the early days of the Footprint, we made big mistakes. For example, we had buttons that admonished people to "reduce your Footprint!" When people saw these protest buttons, I am sure many just thought to themselves, "You go right ahead and show us!" After a lecture I gave in Chile, an astute student asked, "Am I supposed to reduce my Footprint so you and your compatriots can eat more of your famous Swiss chocolate?" That moment was brilliant. It helped me recognize the severe limitations of my own communication. What's in it for others? Clearly we had failed to make a good case. This student's honest and profound question made my trip to Chile worthwhile.

BERT: Yes, she clearly spotted your wagging finger...

MATHIS: Indeed! Another one of our mistakes was calling the world's *per capita* biocapacity "our fair earth share." The self-righteous moralists loved it. But skeptical people who we try to win over reacted with, "Who are you to decide what's fair and what isn't?" And they are absolutely right. What is fair has to be decided collectively, not imposed by me. It's far better to simply describe what is. For instance, that in 2019, the world had 1.6 biologically productive hectares per person. That's an observable statistical average. It's just a fact. We have to describe what is, not mixing it with judgment. This is also the reason why we are separating the National Footprint and Biocapacity Accounts, the descriptive part, from Global Footprint Network, the advocacy part. Those accounts need a separate, independent entity.

BERT: So you keep judging?

MATHIS: Yes, I am human, and I have my views. But I am striving not to mix description with interpretation. I present it like this: Here are facts, and then here is my view why they are relevant. Advocacy, or standing for a world where all can thrive within the means of one planet, is a particular choice, or you could even say a judgment. This underlying vision is the DNA of Global Footprint Network, and also what I want. My goal is to find ever more people who also want it, and who also recognize that this requires that our decisions are consistent with our vision. For that, what really matters is to keep our communication productive and engaging. Footprint communication should make people want to get involved. Judging others or showering them with doom and gloom without providing a sense of possibilities of how to avoid those risks is off-putting. I also need to drop my arrogance, and need to be willing to genuinely listen to other people's perspectives. This is one of my aspirations. I am still a student. But even if we officially ban any language of telling people what to do but privately think about norms they should follow, people will read between the lines. If I say, "We must drastically reduce our consumption of energy," they will hear, "What he's really saying is that life is a lot easier with more energy."

BERT: So you don't address the question of justice and fair distribution?

MATHIS: Naturally the question of justice pops up instantaneously. It is much more effective if the question of justice is raised by our audience. Justice and fairness are a huge motivation for me, but it is a common mistake to confuse motivation with strategy. The point is not to tell others how to reduce their Footprint, but to give them the tools needed for being successful, and help them recognize that their success is more likely if they are one-planet compatible.

BERT: Let's talk about an important concept in your work, overshoot, or the situation where more trees are logged than can regrow, more fish caught than fish stocks can regenerate long-term. How are we to deal with overshoot?

What Does Overshoot Feel Like?

MATHIS: Hardly anybody in the world with money or influence ever experiences overshoot directly. Money papers over the cracks. People may simply no longer vacation in Haiti because it has ceased to be pleasant there. To put it bluntly, people with money experience the gross domestic product, abstract as it is, more directly than they experience overshoot. If the economy is running well, people with high incomes have more purchasing power and with it opportunities. Everything becomes easier, from finding a new job to satisfying some of one's desires to getting a loan. Also the value of one's residence goes up. These are all immediate effects of higher GDP on people with high incomes. As long as the constraints of overshoot tighten more slowly than the urban elite's purchasing power increases, that elite won't notice resource constraints. Even if the economic situation turns turbulent and the gross domestic product shrinks, people with high incomes will barely recognize the signs of overshoot (small exceptions may be the recent water crises in São Paulo or Cape Town that also affected people with money). And when crises hit, many will demand that the government promote more consumption because the party must go on.

BERT: Yes, but some governments also react with austerity. Is this the solution?

MATHIS: Hardly. To me, the fact that governments only consider two options to react to economic crises, stimulus or austerity, is a symptom of the problem. It emphasizes society's focus on income, rather than concern about society's wealth. And with society's wealth, I do not mean the sum of their money, but their ability to produce long-term income. It includes skills, health, trust, and yes, resource security. Therefore, the real question should be: Are we spending our money in a way that builds rather than erodes society's wealth? This should determine how we spend our collective financial resources. But during a crisis, few people have the courage to look into the belly of the beast and ask what is really driving the crisis, and what we need to do differently.

BERT: What would they see in the belly of the beast?

MATHIS: Well, in a finite world, we cannot stabilize an economy long-term by constantly increasing consumption, especially if we are already in a state of overshoot. As long as we live on the liquidation of our natural capital (as is the case with our current economic practices), we undermine long-term security for everyone and everything on Earth. The way we pay for the present by liquidating the future truly fits the definition of a Ponzi scheme. Any other forms of Ponzi schemes are outlawed, only the ecological one we seem to ignore or even encourage.

BERT: Is our money focus blinding us from the ecological Ponzi scheme?

MATHIS: This may well be the case. Remember, money has no real value by itself; it is merely a symbol that allows access to real wealth. Let's unpack societal wealth a little more—our assets that allow us to produce truly sustainable incomes. What is core to true wealth are human capital (skills, labor, health, knowledge), natural and particularly biological capital (resources, waste sinks), and physical capital (houses, factories, railway lines). Our amount of human capital is going up, while that of natural capital is going down. All forms of capital depend on having sufficient natural capital. So instead of liquidating our natural capital, we should invest in it because natural capital will become scarcer and hence ever more important. It would also mean to invest in physical capital whose value does not rest on the liquidation of natural capital (for example, more zero-energy buildings but fewer jumbo jets). We should distinguish more wisely between investments that retain their value in a world of overshoot and those whose value will erode. A good economist ought to carefully watch the state of our natural capital.

BERT: For that we have the Ecological Footprint, which even gives us numbers to work with, such as the global consumption of Earth's biocapacity in 2019: 75%.

MATHIS: Yes, we need such numbers to help decision-makers comprehend the state and trend of the natural world. I am pretty sure astronauts want to monitor the life support systems of their spaceship. Particularly if the crew overstrains the spaceship's life support system, and if there is no other spaceship in sight to rescue, once the support system gets exhausted. Oh, by the way, that's our situation on planet Earth…

BERT: How can your numbers affect the scenario you have just described?

MATHIS: Global figures give an individual person important context, but still little guidance for action. To act, people need more specifics as well. And the numbers have to be relevant to their concerns. That's why we also break the results from the Footprint accounts down for countries, cities, individuals, and products. For example, for countries, cities, or individuals, we can apportion the total Footprint to activity fields—for instance, mobility or food. And those fields can be broken down further—food is split into as many subcategories as data allows for. Such detailed numbers are essential in order to identify big-ticket items. Also, different activity fields require different types of policies. More detailed numbers help device strategies and monitor outcomes.

BERT: Why would this be of interest to people?

MATHIS: A large portion of humanity experiences impacts of ecological overshoot firsthand. Therefore, being able to track those dynamics seems of immediate relevance particularly to most people on the planet. Most of humanity lives in lower-income regions with little biocapacity. In Kenya or Bangladesh, for example, overshoot translates into droughts, food shortages, unemployment, and ultimately social conflicts. Low-income farmers in India cannot simply expand their farms to produce more. Agricultural inputs and technology are expensive, and water is often scarce. As a result, agricultural productivity in many low-income areas does not increase at the same rate as the population does. If these trends are not reversed, the situation will become continuously more distressing.

For many, limits are only real when we hit the wall. But, in fact, there are many early warning signs that when heeded allow us to avoid train wrecks. Overshoot, like financial overdraft, is such a warning sign—it is not the final wall—yet. There are invitations to correct our path, or face the unpleasant consequences. A forest, for example, can be overexploited for a long time—but not beyond a certain point. Water tables fall slowly but steadily. Fishing grounds collapse faster than forests or water supplies. It has already happened repeatedly. People with high income experience such a collapse only if they live in coastal regions that are highly dependent on fishing, as in Newfoundland. City dwellers who get their groceries from the supermarket will still find enough fish in the freezer. They may even find it exotic to eat new species of fish; for them, it's enriching.

BERT: Rather perverse.

MATHIS: The last trawlers will deliver their load to those with the highest purchasing power. As the gap between high- and low-income people is opening further, society's ability to react is further undermined. The reason is that influential decision-makers (who are typically in the high-income segment) are removed from the realities of overshoot. This means that our system loses necessary feedback. If those who steer society are ever more shielded from physical reality, course corrections become unlikely. Overall, the experience of many influential city dwellers is still one of expansion. In times of economic crises, they merely complain that there is "not enough expansion" and demand stimulus packages or deregulation to boost the economy.

BERT: That sounds disastrous. So, it appears overshoot becomes only palpable when things are already harsh, such as when there's a collapse of a stock, or when people get pushed up against the wall and may no longer have the economic means to react.

Causes and Symptoms

MATHIS: People may expect contractions or collapse to be fast, in matters of days or weeks. Financial systems can produce such rapid shocks or even halts, whether it is a Ponzi scheme breaking down, a bankruptcy, a stock market contraction, or a currency devaluation. Contributing resource factors may not be obvious to most, particularly as environmental erosion is a slow grind. Social implications of environmental bottlenecks eventually become more visible: for instance, as disruptions cause financial shocks or diminished opportunities amplify social conflicts like the extended Arab Spring with the ensuing refugee crisis in Europe. If you listen to daily news stories, nobody ever dies from overshoot. Instead, they die from its symptoms: economic breakdowns, poverty, wars, pandemics, natural disasters. The issue is whether sufficient people in mainstream society will recognize the underlying dynamics. Or will we continue to feel surprised by symptoms?

BERT: Overshoot is a pretty abstract concept.

Reducing Complexity

MATHIS: To some degree, yes. But the great strength of Footprint accounting is its ability to reduce everything to a common denominator. It's helping to bring back physical reality into social sciences.

BERT: In the past, decision-making was easier: we built a road and were done with it. Today we have to consider biodiversity, water levels, erosion, food prices, carbon dioxide. How is it all interconnected?

MATHIS: Overshoot paints the large picture; it provides a context that weaves all these aspects together. Countries that understand and manage the reality of operating in a one-planet context are in a far better position to navigate the challenges of the 21st century. Those who prefer to stay in their illusion bubble, however, will keep getting crushed in continuous battle. This should be of interest to economists, too. For the growing natural capital constraints will shrink economic opportunities, possibly driving stagflation, certainly push us into vicious economic down cycles. Economically, overshoot translates into daily necessities such as energy and food becoming more expensive while many traditional investments in physical capital—for example, houses and other resource-devouring objects of value, such as industrial plants or airports—dropping enormously in value. Continuing to "grow" during overshoot is like burning the candle from both ends. It will destabilize many markets. At that point, the cost of resources will drop again, but so will people's purchasing power.

Where Does the Goal Come From?

BERT: With Footprint accounts, we have a metric to identify and track overshoot. But it doesn't tell us how to respond. What does it take to end overshoot, according to Global Footprint Network?

MATHIS: At Global Footprint Network we deliberately limit ourselves to one decisive question—how much biocapacity do we use, and how much do we have? We focus on making our answers as accurate as possible, the inquiry as relevant as we can, and our interpretations of the results as empowering as our creativity allows. This is in service of pursuing the goal of helping humanity to maneuver itself out of overshoot, or to end overshoot by design, rather than just waiting for the otherwise inevitable disaster to do the trick. This is the foundation of our work. We see overshoot as the mother of all trials and tribulations humanity will face this century. Mainstream society vastly underestimates the significance of overshoot. The number of people overshoot will affect, the likelihood of its occurrence, and the scale of its impact turn it into the highest risk of any that humanity faces. The good news is that this dilemma can be approached constructively. Overshoot is not like an unavoidable asteroid that suddenly hits Earth. We understand the phenomenon, we can measure it, and we can do something about it.

BERT: What should we be doing?

MATHIS: Because collectively, we are so far away from wanting to address overshoot, Global Footprint Network is dedicated to work on two ingredients to have decision-making reflect the context of overshoot. One effort is to provide neutral, trusted, robust accounts— soon through an independent body. The other path is ramping up meaningful engagement with the issue. Our contribution is to make ending overshoot more prominent in the public's agenda. For instance, with Earth Overshoot Day, we now generate over three billion media impressions per year in over 100 countries. This is the number of people who have access to the overshoot message on their media platforms.[4] They may not all look, but they could. Media impressions are still far from turning insight into action. Still creating the language and clarity of the challenge is a necessary ingredient for this needed transformation.

BERT: And then?

MATHIS: Of course, the world needs much more to get out of overshoot than what we as one organization can offer. A widely embraced and cherished vision of what one-planet compatible living could look like is necessary as well as concrete demonstration examples that show the way. And a shift in our development theories so they become compatible with our one-planet context. Our contribution is to outline pathways through Earth Overshoot Day focus on #MoveTheDate—the idea that if we pushed that date four to five days into the future every year, humanity would be back to one planet before 2050. Cutting CO_2 emissions in half would buy us 89 days. This seems very feasible.[5]

BERT: How do you suggest we get there?

MATHIS: With willpower. I love the statement attributed to Antoine de Saint-Exupéry, cherished author of *The Little Prince*: "If you want to build a ship, don't drum up people together to collect wood and don't assign them tasks and work, but rather teach them to long for the endless immensity of the sea." That's what I believe will get us there and what needs to be done. If decision-makers come and ask us how they can reduce their biocapacity deficit, we can offer them a pile of references, reports, and examples. But first they must want them. We believe, though, that people can't make much progress if they don't have robust measurements. Hence it is our priority to advance a well-functioning, scientific, and intelligible tool to quantitatively describe overshoot. In our eyes, having a reliable accounting system for our resources is an essential condition for reaching our primary goal of ending overshoot. And we hope that this tool, together with the reasons we give for reducing biocapacity deficits, will help governments of cities and countries and investors to recognize their self-interest in recognizing the one-planet context. I hope this will catalyze necessary change. And to be honest: our goal implies a norm and makes the value judgment that we'd be better off without global overshoot. We'd love to do more engagement than just Earth Overshoot Day campaigns. They are just an opening gate to our "engagement funnel."

BERT: Could the Footprint method also be used—to be brutally blunt—to map out or maintain eco-dictatorships?

MATHIS: Theoretically, yes. Overshoot is the decisive factor of the 21st century. That's why we have to confront it—in dictatorships, autocracies, democracies, or any form of government. Accounting does not favor dictatorship. Rather, an unbiased, rigorous accounting system increases the transparency of decisions and therefore serves as an important instrument of a successful, participatory democracy. I certainly do not wish for eco-dictatorship. Authoritarian rules become more likely if there is no accounting of resources, no transparency, no negotiating, and no polluter-pays principle. With the help of Footprint accounts, people can evaluate governments and their decisions. And those are fundamental dimensions of a functioning democracy, particularly if it wants to survive and thrive in an ecologically constrained world. The future of humanity depends on such transparency, for better or for worse. Either way, we don't want the rigor of our good accounting method distorted, watered down, or manipulated by special interests. Protecting the clarity of this accounting approach has been a central task of Global Footprint Network, one that will also be advanced by a new entity looking after the National Footprint and Biocapacity Accounts.

Try out Any Potential Solutions!

BERT: What is your take on "fair fly"—the concept that people pay a certain amount to compensate for their air travel, an amount that can then be used, for example, to redevelop forest areas?[6] It's a way of neutralizing one's air miles, so to speak. But is it not a form of selling absolution?

MATHIS: Humanity is in such a tight spot that we have to try out anything we can come up with to reduce our resource dependence. Until now, nature's service of removing carbon dioxide from the atmosphere did not cost us anything. "Fair fly" offers are at least the beginnings of a voluntary marketplace. But the price of these absolutions is far too low, and the marketplace too small. A functioning market would generate more money to regenerate natural capital. Besides, the amount of people's purchasing power that they spend on these compensations can then no longer be spent elsewhere. Also, if the price is high enough, I will fly less, and so will many others. We therefore should give initiatives such as "fair fly" a try to find out what works and what doesn't. If nothing else, compensation payments create economic opportunities. If we take our situation seriously, we will have to introduce such initiatives on a larger scale. Voluntary abstention from consumption will not suffice. And there is cost associated with calling new compensation payments such as "fair fly" absolutions or indulgences. It may primarily stimulate cynicism in the process. Once cynicism becomes an excuse for doing little or nothing, the situation becomes dangerous. Inaction has already grave consequences today. Is it an exaggeration to call our collective inaction a crime against humanity?

Who Will Get the Fish?

BERT: Watch out. Now you are about to start moralizing…Let's take another look at overshoot but this time from a different angle. How much unspoiled nature, how much biodiversity does the planet need?

MATHIS: The 1987 Brundtland Report, *Our Common Future*, proposed to set aside 12% of the world's land as nature reserves, a number that was politically motivated.[7] Today, we are globally indeed somewhere in that range with our nature preservation and landscape protection. Unfortunately, many of the protected areas are marginal and represent only a distinctly smaller portion of the world's biocapacity. Biologists have estimated how much biocapacity we would need in order to sustain certain parts of our biodiversity. If we were to concentrate on the most diverse hotspots, we could sustain much biodiversity with quite limited areas. Unfortunately, many of these hotspots, or most valuable regions, are already heavily used by people. That's why areas beyond the hotspots are needed for meaningful ecosystem and species protection. Professor E. O. Wilson summarized his dream in 2003, proclaiming to leave half Earth under protection as our most important inheritance and investment in biodiversity. Initially, the idea was just a footnote in his book *The Future of Life*. In 2016, Wilson doubled up on his idea in his book with this very title, *Half-Earth: Our Planet's Fight for Life*; he also started his latest organization, the Half-Earth Project. There is also a coalition of conservation organizations called "Nature Needs Half" pursuing the same goal.[8] Sadly, humanity does not—yet—follow this advice. When explaining biodiversity loss, some people use the analogy of a plane whose rivets are being pulled out. How far can it travel before it breaks apart? Nature is considerably more tenacious than planes. As a result, we don't get enough feedback even though we are losing an extraordinary number of rivets.

BERT: What will happen next?

MATHIS: Though it will be sad for the planet's biodiversity, human beings will likely be able to survive in a considerably impoverished ecosystem precisely because nature is so tenacious. Just look at the Netherlands, for example, with an entirely artificial landscape. Yet it keeps chugging along (also thanks to enormous imports that feed their pigs and themselves). That paradox is that the human experience does seem to get better by many measurable outcomes, as Steve Pinker points out.[9] Just that we are also massively depleting the life-support system on which we depend, a foundational threat Steve Pinker vastly underestimates. By the time we experience the feedback and are directly threatened by lack of biocapacity, the world will have become quite desolate. And we will accept it, just as today we accept depleted watersheds, widespread eroded areas in the Mediterranean basin, ever more diminished coral reefs. In the countryside, we are losing beetles and beautiful birds; we are losing primates and possibly the rhinoceros, while humanity keeps focusing on expanding its enterprise. The question is: Who will get the last fish? The sea lion or humans? We can't both get it. Humanity is one of the most adaptable species and is able to survive in all climate zones. Surely we won't be the first to go extinct. Not as a species. But that does not mean we can maintain the size of our current population.

BERT: In essence, the Footprint is an indicator, with strengths and weaknesses. A weakness of another indicator: the gross domestic product counts (for example) every car accident as an event that will grow the economy.

MATHIS: It is crucial to have clarity about the question an indicator answers. Problems arise when an indicator is seen as answering a different question. The gross domestic product (or GDP) answers how much monetary value we see added in a year to a country's economy.

GDP does this amazingly well. A more fitting name for it would be gross domestic market transactions. But, regardless of its name, it becomes misguided to use this indicator as a proxy for progress. It is misguided because GDP does not track many aspects of progress, for instance, whether we are healthier, happier, or safer. These are, by GDP's very definition, not questions the gross domestic product addresses. Yet by following GDP as the overarching metric, we behave as if it measures everything. It is true that the expenses caused by a car accident are interpreted as an increase of the gross domestic product. Yet these expenses are defensive costs that are not productive in the long term. They are costs (like forest fires, collapsed bridges, or crushed cars) that can undermine the economy's productivity. Such costs and the associated short-term spending erode the economy's potential for generating well-being and income in the future. In effect, maintaining the gross domestic product will become more difficult in a year following too many defensive costs. And obviously, these expenses do not make us any happier, either.

Illustration: Phil Testemale

BERT: So let's not be unfair toward the gross domestic product.

MATHIS: Indeed, the gross domestic product has been brilliantly designed for what it is intended. It keeps account of final sales, just as the Footprint keeps account of biocapacity. As the Ecological Footprint is limited to its question, so is GDP. GDP is blind toward any value creation outside market transactions, such as non-monetarized labor in childcare or volunteer work. Neither does it capture value depletion or wealth loss that are not compensated. Draining our natural assets by $200 while making $50 of sales is still counted by GDP as a $50 gain, not a $150 loss. For instance, GDP doesn't measure the use of groundwater because groundwater comes "free of charge."

We need to know far more than GDP if we want to make good decisions for our communities. In essence: an indicator is powerful in the realm of its research question. That's where GDP or Footprint accounting score high. Unfortunately, there are plenty of indicators out there that do not even have a clear quantifiable research question. They are arbitrary indices adding up things according to a made-up scoring approach. They should be banned. But outside the realm of its research question, any indicator is pretty useless. In my view, Footprint accounts answer a more relevant question about sustainability than GDP. But it is still useful to have GDP as an indicator, just not as the prime one.

What the Footprint Can Do—And What It Can't

BERT: You explained the limits of GDP. Where are we asking too much of Footprint accounting?

MATHIS: Footprint accounts merely answer how much of Earth's regenerative capacity is being consumed by human activity. They communicate about the competing human uses of the biosphere; they show that humanity is currently operating in global overshoot, and they can show which activity contributes how much to the overall total. However, we'd be off track if we managed all of our economy and policies exclusively according to Footprint accounting. After all, our ultimate goal is to protect everybody's quality of life. A necessary precondition for meeting this goal is not consuming more than Earth can regenerate. But there are other requirements as well. For example, we need to make sure we keep diseases under control. HIV/AIDS or malaria, for example, have no Footprint, but that doesn't make them unimportant. Too much traffic noise is no fun either. Ugly cities or landscapes are equally undesirable. Injustice is destructive and inhumane. Unequal access to resources can be measured by the

Footprint, but that is only one aspect of injustice. We need additional indicators.

BERT: Global Footprint Network sees itself as upholding its methodology and looking after its standards. The network is managed by a team in an open, transparent process. Do objections against the Footprint exist that you find worth considering?

Objections

MATHIS: I do. One is structural—the conflict of interest between being accountants and giving advice. In the finance world, these functions are separated. In response we are working to separate these functions by making the National Footprint and Biocapacity Accounts independent. There's also a whole long list of methodological improvements that should be researched. There are infinite examples: How to track trade more accurately, given noise in trade data? How to distinguish between hardwoods and softwoods? How to more accurately track the productivity of fisheries against harvest rates? Or how to account for nuclear energy? This last one is more complex as the way to capture nuclear demand on biocapacity is not straightforward. The most relevant concerns regarding nuclear energy may be other dimensions such as military proliferation (including the potential for dirty bombs), costs, operational risks, or long-term storage of waste. Still, there is a link between nuclear accidents and biocapacity: Calculations show that the Fukushima accident claimed significant amounts of biocapacity. The exclusion zone around the damaged reactors represents about 1% of Japan's biocapacity and may remain closed off for a hundred years or longer. Put differently, that one nuclear accident occupied at least 100% of Japan's biocapacity for at least a year. That's not a trivial amount.[10]

Another open Footprint question is the relationship between biocapacity and biodiversity. Or organic versus industrial farming. Which one has a larger Footprint? I am not sure. Superficially, people may assume organic farming has a larger Footprint than conventional, as annual yields are sometimes lower than those of intense industrial farms. But organic farming may score better on Footprint because of less resource-intensive inputs, and better maintenance of long-term yields due to soil health. And organic agriculture has additional (non-Footprint) benefits because it uses no synthetic pesticides, is more biodiversity friendly, and often kinder with animals. Lower short-term yields are also a price well worth to secure yields that continue in the long run. And a commitment to organic farming does not mean we'll starve, either: we can eat less meat and more vegetables, something that's undoubtedly healthier, too. Footprint is not the be-all and end-all. It simply addresses one distinct and, as I believe, essential issue: if basic quantitative goals are not met, it is not possible to scale quality. We can have a wonderfully cared-for forest in one corner, but if overall timber demand exceeds what the forest can produce, protecting one piece of the forest just moves the excess demand to the less protected parts of the forest. Quality can only be extended to the whole if basic quantitative conditions are met.

BERT: In the context of fossil energy and climate change, the Footprint focuses exclusively on carbon dioxide. But there are other greenhouse gases, such as methane. What about them?

MATHIS: This is another issue for the research agenda. We ought to integrate those other greenhouse gases into the accounting method. The research challenge is how to allocate the pressure from those gases accurately to final consumption. Including those gases into Ecological Footprint accounts would increase their estimates of overshoot and biocapacity deficits. Not including them makes the accounts less accurate, but at least it contributes to not exaggerating the biocapacity deficit situation.

BERT: One objection is that the carbon Footprint is merely hypothetical and not real. How do you see it?

MATHIS: We can't see or smell carbon dioxide emissions, but they consist of matter and are measurable. They form a physical flow of waste that can be absorbed by ecosystems. Humanity could allocate more biocapacity to absorbing carbon dioxide currently released by our burning of fossil energy, but this would subtract from capacity for producing potatoes, milk, or wood (yes, there are also possibilities to soften the trade-off through better farm management—but it is still a trade-off). So there is real competition for biocapacity. Carbon Footprints compete with forest product Footprints and food Footprints. and so on. That's what Ecological Footprint accounting measures. Also, the fact that there are companies and organizations offering carbon sequestration through reforestation indicates that indeed there is a link between the land and the atmosphere. The issue is, even as the Paris Climate Agreement confirmed, not emissions, but net emissions. We are not dealing with just the atmosphere.

Including the carbon Footprint as one of the competing pressures does not mean we claim, as some critics maintain, that the climate problem can be solved through afforestation. On the contrary, our numbers show that the human addition of greenhouse gases is so massive that nature's ecological services are insufficient if we want all the other services from nature as well: land for food, wood, or cities. Afforestation and better land management can make an important contribution, but are not by themselves *the* solution.

As always, the devil is in the details. For instance, no reliable data exist at this point that documents whether the regenerative capacity of forests around the world is growing or shrinking. With climate change, forests can become more prone to forest fires or new pests. Both can lead to huge productivity losses while forests adapt to the new climatic conditions. Given the limited data, the real situation is probably quite a bit tighter than Footprint accounts reveal. Still, calling the carbon Footprint "hypothetical" just because it is much larger than the land allocated to carbon sequestration is plain silly. A timber Footprint larger than the forest is not hypothetical either. In both cases, the excess of Footprint compared to biocapacity is overshoot. That's why we reply to our critics that what they call "hypothetical" ecologists call overshoot. It is, unfortunately, real.

How Precise Are the Footprint Numbers?

BERT: Another objection to Footprint is that it is a highly aggregated indicator and therefore fuzzy in the details.

MATHIS: Footprint is an aggregate indicator much like the gross domestic product. Both are accounting systems; both answer a specific, consistent, and empirically observable question. We can interpret their results in their respective contexts and scale them down for specific sectors. For example, the gross domestic product lets us see which economic sector generates what added value. In a similar way, Footprint accounting is able to describe in highly specific terms the consumption of nature at the national level. The practice of Footprint is not yet as advanced as that of the gross domestic product, which has been developed for over a century and which many countries introduced more than 60 years ago as a public measuring tool. But even that tool is constantly being fine-tuned, just like Footprint. Also Ecological Footprint accounts are limited by the data. Describing the Ecological Footprint and biocapacity of a country with up to 15,000 data points may sound like a lot, but it gives the National Footprint and Biocapacity Accounts still a meager resolution.

BERT: How helpful are Footprint numbers when it comes to planning details, for example for cities?

MATHIS: When the BedZED project in London was built in the early 2000s, the goal was to stay within the now 1.6 global hectares that are available for every person globally. That goal was not quite reached. Had we had the Footprint methodology in today's form then, we might have been able to produce relatively elaborate measurements of the project. Still a few studies were done to assess its performance, and it is clearly much better than its neighbor developments. Possibly by a factor of two for the residents' entire Footprint.[11]

The benefit of Footprint accounting is that it is consistent and describes not only parts of our consumption but the whole picture. With the help of Footprint accounting, we can determine for communities or even for parts of a city exactly how much nature is needed for cars, public transit, the heating of buildings, or meat consumption. A consistent framework like the Footprint accounts are needed to organize and compare data, from the national all the way to the project level. The Footprint draws an absolute line, similar to the Plimsoll line on the side of a ship. It shows how much cargo we can load on the ship without sinking it.[12]

Is the Footprint Fair?

BERT: It's bad luck for the Netherlands that the country is so densely populated. Russia, by contrast —the only large power, besides Brazil, that has sufficient biocapacity to cover its Footprint demand—has been endowed with great natural wealth. Is it historical luck? Is it unfair?

MATHIS: Fair or unfair, when the World Bank presents its figures for gross domestic products, there are no demonstrations in the streets of Washington either. No one protests against GDP accounting because it documents the unfair reality that Ethiopians receive on average far less income than Americans. The numbers simply describe what is. I am not legitimizing those differences. But they are a description much like a photo is a description. And yes, it is unfair that one person has a big farm, another one a small farm, and a third one no farm at all. We can describe it, though.

BERT: Where will the Footprint method be ten or twenty years from now?

MATHIS: If we have success in helping our audiences see the value of understanding biocapacity, there will be tomes of awfully boring scientific studies and standards, and there will be conferences for the statistical offices that will prescribe how exactly Footprint accounting is to be further improved and developed. The more standardized the method will become, the more difficult and expensive it will be to change and adapt it. Footprint numbers might get used as context for competitiveness plans and infrastructure decisions. They may also inform international negotiations or even be considered as basis for compensation payments. In any of those applications, it is important that the accounts are reliable, trusted, and neutral.

Alternative Methods?

BERT: Are you aware of any other indicator competing with the Footprint?

MATHIS: I'd be happy if there were. Of course there are related measures, which is wonderful. For example, there is the effort of

Planetary Boundaries spearheaded by the Stockholm Resilience Institute and now many others. It uses nine categories to compare human consumption with nature's metabolism. Planetary Boundaries' goal is to examine environmental impacts that could take our system to a tipping point.[13] That's not exactly what we measure, because human demand can go over the threshold of regeneration before it reaches a tipping point. Still, the Planetary Boundaries analysis is highly complementary to Footprint accounting. The nine boundaries highlight aspects or conditions that enable (or thwart) the ultimate outcome: biocapacity. Their analysis also independently confirms our results: we are indeed in a state of overshoot. Planetary Boundaries is hence a great support for our work. Otherwise, there is little "competition." There are other metrics of consumption, though hardly any that compare consumption with nature's regeneration. We need to measure both sides: demand and availability. Financial accounts that only look at either income or expenditure are not really that useful. It is contrasting the two that generates meaning. The same is true for Footprint accounts.

BERT: So what are the alternatives?

MATHIS: There are comparable methods, such as Net Primary Productivity (NPP) and the human appropriation thereof (HANPP).[14] That method asks: How much biomass can an ecosystem produce? How much of that amount is harvested by humans? The method is useful for a range of applications. For example, it lets us examine the intensity of our consumption, or the "depth" of our Footprint. This complements the Ecological Footprint in that it tells us more about the intensity of use, rather than the special extension. But the method cannot determine with sharpness how much of this productivity can be consumed and where the sustainable limits exactly are. NPP is thus a related, complementary method but does not answer the same question as Footprint accounting. Then there are other crucial metrics that we, too, draw on, such as statistics about physical mass flows. If we restrict our statistics to the flow of money, we blind ourselves. We need information about mass flows, both with an eye on production and on trade. Footprint accounts then convert those flows into how much nature is needed to maintain them. Luckily data on international mass flows as well as the development of Life-Cycle Assessments for the measurement for mass flows associated with products and processes are becoming ever stronger. Our own analyses depend on these statistics.

BERT: Let's take a look at the Footprint map of ecological debtors and creditors. If a country has much biocapacity, it does not automatically mean that this natural wealth will benefit its population. On the contrary. Some countries on the African continent are blessed by natural abundance and yet their people starve. Does such a situation enter into the Footprint calculations?

MATHIS: Footprint accounts just map availability and use, not how well biocapacity is used. Having sufficient biocapacity available, whether locally or through one's purchasing power, is a central physical condition for enabling economic activities. The social key ingredient, parallel to biocapacity, is trust. Without it, collaboration, innovation, market exchanges become difficult. High levels of trust in society, reflected in stable institutions and rule of law, enable amazing things to happen. Congo is a staggering example of enormous biocapacity but low levels of societal trust. In this case, biocapacity does not translate into human well-being. Yet that biocapacity may get used by others—Rwanda, without its easy access to Congo's (undefended) natural resources, would not be able to operate as well as it does.[15]

Let me emphasize, being an ecological creditor is not a guarantee of success. There are extreme cases (such as Afghanistan, Chad, Somalia, or Sudan) where violence or civil war make parts of a country inaccessible. These countries then appear as ecological creditors because they have more capacity than they consume, while in reality it is tragedies that limit their already minimal Ecological Footprints.

BERT: What can politicians do with your Footprint numbers?

MATHIS: I guess it only works for those who see overshoot as a risk. For those who do not see it as a prime risk, my role becomes to challenge their perception. For those who recognize the significance of the risk, I would start exploring with them how Footprint and biocapacity trends affect their country's or city's competitiveness. This would lead to identifying their country's (region's or city's) Ecological Footprint and biocapacity goal. How much biocapacity do they want to use by when? Too much is a risk, and too little is uncomfortable. Once this goal is set, and it truly reflects what they want (not just something they say to look good), then the task becomes straightforward. Establish performance benchmarks. Such benchmarks are quite straightforward: just divide the total Ecological Footprint savings they want to achieve with all the budget they have available over the time period. Every dollar they spent that produces less than the calculated reduction per dollar becomes a liability. Every project that underperforms in increasing their constituency's resource security has to be compensated by a project that overperforms if they want to stay on track. It is really quite simple.

Rules of the Game in the 21st Century

BERT: Are there countries that already feel the impacts of overshoot? MATHIS: In Haiti or Darfur, the resource situation is already so tight that the economies have practically run out of steam. What nature generates is no longer sufficient to properly feed the populations that live there. And most people there do not have the income to import what they lack (of course, even there you will find some people with high income who can get what they need, but on average they cannot). Niger is in a similar situation. Socially, this is turning into a humanitarian disaster characterized by unspeakably difficult and destructive living conditions. Yet these areas of the world don't play any significant role on Wall Street and hardly register with people in socio-economic bubbles, like yours and mine. We urbanites who live in the upper percentage points of the global income pyramid can easily ignore those countries. Things could change, though. Increasing resource constraints will affect growing numbers of people, and political instabilities can become hot spots of international conflict. The Arab Spring and later ISIS are all symptoms of inhumanly tight resource situations.

At the same time, high-income countries are not immune either from the implications of global overshoot. The old rules of the game of infinite cheap access to resources, whether through colonialism, unequal terms of trade, or systemic externalities, are starting to crumble. As a result, access to resources is becoming an economic factor. For example, cities or states who want to avoid becoming marginalized need to contain and reverse their excessive resource dependence. Every decision—whether it is building roads, housing, or dams; planning city districts; designing power supply, or any other kind of investment—will gain in power and relevance if it also responds to how it will enhance (or weaken) the economy's resource security.

BERT: An example, please?

MATHIS: In Oakland, California, the then mayor Jerry Brown (who later became Governor of California) challenged suburbanization trends by advocating for "vertical suburbs." He recognized that the value of Oakland would increase through densification, integrated zoning, and ensuing vitality. When people live where they work, and cities are livable and vivacious, they are also safer. Such cities thrive on proximity, and can shed their car dependence. It does not mean that densities should be extreme: skyscrapers devour excessive energy—the optimum is more in the neighborhood of four to seven floors max. Jerry Brown's vision worked, in concert with many progressive community groups that advocated for bicycle transportation, integrated and socially just development efforts, and who celebrated the creative spirit of Oakland—now summarized in the city's informal tagline: "oaklandish."

But let me also be honest: The speed and scale of Oakland's transformation is still lagging enormously behind what one-planet compatibility truly requires. The city continues to approve developments that are incompatible with a fossil-fuel-free future. For instance, a current building boom is generating new housing with insufficient energy efficiency, lacking ways to fully harness solar power, and excessive car amenities.

Oakland, the American home of Global Footprint Network and where I live, is dear to my heart. But imagine if this amazing city could be turned into a kind of Paris, with a higher population density through four-to-seven-story buildings and boosted high quality of life in an extended downtown and Oakland's many vibrant neighborhood nodes. Mobility could be largely made possible by two-wheelers—public and private bicycles and scooters, including electric ones—walking, complemented by public transportation. In addition, there could be electric rickshaws, perhaps also some electric taxis. The moderate climate of Oakland would allow for buildings that could be energy positive. Oakland would be far better than Paris, because we could take advantage of the newest and most efficient transportation and housing technologies. Still Paris is a valid example because it was built before the Oil Age. By necessity, it had to be efficient. Energy was expensive: it took much human labor and energy to get an extra unit of energy.

Maybe Paris is not the best example either because I doubt that cities with more than a million inhabitants will function well in the future. Supply routes will simply be too long, as their Footprint will have to extend too far.

Efficiency Is No Universal Remedy

BERT: Efficiency has become some kind of magic word. But Footprint numbers show that overconsumption of resources has increased in spite of efficiency enhancements. More efficiency lets us get more goods and services from a given slice of nature, but it cannot increase how much nature is available.

MATHIS: Efficiency alone seldom reduces resource consumption. Particularly if efficiency gains are profitable, they encourage doing more, typically leading to increased consumption, as economist Stanley Jevons pointed out over 150 years ago.[16] More efficient long-range jets now fly longer routes, and fly more often and at lower prices. Profits from efficiency gains are rarely used to protect resources. However, resources could be protected if efficiency enhancements were combined with an ecological tax shift. Bill Rees, with whom I developed the Ecological Footprint in the 1990s, once drew on a piece of paper a ship about to sink. Then he drew five of those heavy Hummers on top. What's about to happen? Then he drew the same ship with five Priuses on deck. What would happen now? Either way, the ship sinks, he said with a smirk. Efficiency that leads to more consumption won't save us. But if employed in conjunction with meaningful resource security policies, these technological opportunities could do wonders.

Reality Check

BERT: Ending overshoot—how's that supposed to happen anyway? You developed three scenarios, each following the logic of the Footprint. But the world is complicated. What will the scenarios get us?
MATHIS: Recognizing the reality of biocapacity and tracking historical Footprint trends adds value to conventional scenario approaches. Often scenarios are driven by exploring "what key actors want." Scenario modelers rarely question whether fulfilling these wishes is ecologically even possible. Footprint accounting enables users to ground truth scenarios against physical reality, by having to spell out how much biocapacity they assume their scenario will actually require.
BERT: According to the Footprint's logic, there are five approaches to correcting the imbalance between biocapacity supply and demand, such as reducing population size or enhancing efficiency. Another one would be growing the supply of biocapacity. But is that still a reasonable option on our planet, and if so, could it be done at a level that would make a real difference?
MATHIS: I am skeptical about any claim that biocapacity could grow faster than unrestrained human consumption. It was the idea behind the so-called Green Revolution, and for a while it worked. But in the past decades, the growth of agricultural production has slowed down. Frequently the Green Revolution had side effects, such as a loss of biodiversity, pollution, or pesticide poisoning. Without the Green Revolution, we might be in an even bigger mess. But the Green Revolution should have gone hand in hand with demand management.

We missed that chance. In short, today we need all factors and have to try out all approaches. Particularly because today we need to not only expand food production but also phase out fossil fuel use. Footprint accounting does not provide a catalogue of solutions; rather it is a yardstick to measure what is and to compare what different approaches can achieve.

Collapse

BERT: Why haven't you developed a scenario for ecological collapse?
MATHIS: Because we don't want to forecast a collapse. We want to show ways to prevent a collapse. And we want to warn against scenarios that would drive us into collapse, such as, for example, the combination of the United Nations' conservative scenarios. They are simply not realistic, not even if we assume there will be no negative surprises. They are even less realistic if we were to reach thresholds where ecosystems experience tipping points—if large parts of the Amazon dry out, if the permafrost thaws and emits large amounts of methane, if the polar ice caps melt, or if the Gulf Stream ceases—if any of these events were to occur, then the gap between biocapacity and human Footprint would widen even further.
BERT: I have the impression that you avoid the issue of the collapse. Why is that?
MATHIS: We have too few data to know how many years of planetary debt we can accumulate before an ecological trigger event. It may never be a fast collapse, but rather continuous depletion creeping up on us, distributed unevenly across the world. A fishing stock collapses; a region experiences an acute water shortage during a drawn-out drought; forests disappear and local populations lose access to wood for fuel; soils erode or become too salty. Economically, such situations translate into loss of opportunities, and socially manifest in society through increased conflict and discontent. In hindsight, we can now better grasp what happened in Cuba, North Korea, or Rwanda. Why not use this insight as foresight?
BERT: You get around a lot. What's the international response to the Footprint?

Footprint Variants

MATHIS: The response varies widely and is at times paradoxical. In the UK, for example, Anglican moralism oozes through political discourse. You can see this in both climate discussions or through the rapid uptake of Footprint ideas in the UK. People loudly moralize, the papers are busy reporting, but in the end not much happens. It's all big finger-wagging about moral "shoulds," but no connection is made to economic necessities. Because of the moralism, and the moral pressure, most government administrations are stuck in the misinformed belief that there is no clear economic benefit or economic necessity to react to physical constraints. Administrations put efforts into commissioning reports to legitimize inaction. This makes them "look good" at low cost. That's how the UK government has dealt with the Footprint challenge. At their detriment, I would argue.

The Scottish government, in contrast, has shown more proactive interest—their Environmental Protection Agency even embraced "one planet prosperity" as their regulatory framework, inspired by Footprint accounting. The Scottish government is far more aggressive on decarbonization than the UK as a whole, and their current chief economist recognizes the resource security challenge as a key driver for long-term economic resilience.

BERT: And how do Arab countries respond, which, after all, tend to have big Footprints?

MATHIS: Global Footprint Network has been engaged with the United Arab Emirates for over ten years. As one result, the Ecological Footprint has become a KPI (Key Performance Indicator) of the country, as approved by the prime minister's cabinet.[17] The country's interest surprised even us. The government recognizes climate change and resource constraints as real. The UAE government knows that the country has to refocus its economy eventually. Instead of spending all its oil revenues, the government invests some of them in ways designed to maintain the value of the oil earnings, so the investments can eventually generate income in the post-fossil-fuels age. Some of those investments are channeled into building the country's infrastructure.

However, the government has begun to realize that some of that newly created infrastructure is enormously dependent on oil, and therefore not effective in creating an alternative to oil revenues. In a fossil-fuels-free future, these structures will lose their value if they cannot be radically retrofitted. Tall glass high-rises without solar shading devices, erected right in the desert, are to all intents and purposes solar panels doing double-duty as houses. Of course they are not meant to be solar panels; that's an unintended consequence of architects trying to reproduce Manhattan high-rises in Dubai. As a result of this inappropriate architecture, an enormous amount of energy is needed to cool down these towers to make them livable. It's rather absurd, and the government is realizing that. Cities in the United Arab Emirates have such high energy consumption that the Abu Dhabi government has now opted for a few nuclear power plants (and higher efficiency standards). At the same time, Dubai is building the largest photovoltaic plants with the lowest per-kiloWatt-hour costs in the world. Also, the United Arab Emirates is also the first of the Gulf states to open a large investment fund for renewable energies: The Masdar Fund (also known as the Abu Dhabi Future Energy Company) was established in 2006.[18] One of its projects, Masdar City, is experimenting with super-efficient housing infrastructure that fits the local climate.[19] Overall to my eyes, their progress is hesitant at best, and they have not been able to convince local residents that living in Masdar-like housing is not only more energy efficient but also more pleasant than living in spreading, energy-hogging houses.

With all that, is the UAE on the right track? I'd still say: far too little, far too slow.

Mathis's Dream

BERT: You once said that your and your colleagues' biggest dream is for the United Nations to take over the Ecological Footprint. If your dream were to come true, what would happen ten years from now?
MATHIS: One step in this direction is to develop the National Footprint and Biocapacity Accounts as an independent organization. Hopefully it will become a UN venture at one point…We can emulate

GDP's success. It was able to rise because of the challenging circumstances national governments were in. During World War II, the US government used the gross domestic product for the first time ever in a systematic way. After World War II, financial capital was tight and human hardship grave. But the US wanted to know which was the bigger risk: losing the war militarily or economically. After the war, GDP-led considerations helped bring about the Marshall Plan. And to manage it responsibly, the gross domestic product was needed as a measuring tool. It also allowed for greater transparency (something investors appreciated) as well as fair taxation. This opened the broader success of the metric as it was introduced by many countries, until eventually the United Nations became the standard-bearer for GDP accounting. Ever since, the UN played a leading role in improving and standardizing GDP accounting and applying it worldwide.

Now we realize that our ecological capital is at least as tight as our financial capital, if not more so. This is why we need a similar measuring tool for our ecological capital. For the United Nations to play a similar role as it did for the gross domestic product is therefore one of my big dreams. If each individual country uses Footprint accounting only for itself, the results will not be comparable. Most importantly, countries also need to understand the resource performance of its trading partners, as all economies are materially interwoven. Ultimately, we all live in the same biosphere. Also, I freely admit: if embraced by the UN, the National Footprint and Biocapacity Accounts would need to be massively improved. For this, we need to find countries courageous enough to become trailblazers, who try out the method, press ahead with improvements, and in turn make a better ecological accounting system available to everyone else. This is part of the new effort Global Footprint Network and York University are establishing. Our argument is that courageous countries will have an advantage because they will be the ones who have had a longer time window in which to better prepare themselves for the 21st century.

BERT: Assuming it keeps developing the way you hope…where will Footprint be in 10 or 20 years?

MATHIS: I hope one day we will no longer need Footprint accounting. The ideal world is not a Footprint world. If we take the Footprint calculations to heart and act accordingly, the world of the future will be better than today's. Footprint is a tool for transformation in that it reveals how much we undervalue nature today. Footprint accounting enables us to measure the physical value of our biological capital. Understanding this issue is important not only for us but also, I hope, for all the other animal and plant species with whom we share the planet. Perhaps one day we will even manage to stay well within the constraints of the planet's biological capacity. Then we will realize that life will be not only much more stable and secure but also more satisfying and thrilling.

Acknowledgments:
Who Is Powering All This?

It would be easy if it only took a village to raise a metric like the Footprint. I (Mathis) am thrilled to recognize many collaborators, and petrified about having left out so many more. Deepest apologies to the latter, and many thanks to all.

After the early days with my PhD supervisor and now treasured friend Professor William (Bill) E. Rees at the University of British Columbia, and the fabulous colleagues at the UBC Task Force on Healthy and Sustainable Communities under whose roof I had the privilege to operate from 1991 to 1994, many more started to contribute to the adventure. Too many to name. But in the second phase of the work at the Universidad Anáhuac de Xalapa in Mexico, three people stand out: Anabel Suárez Guerrero and Alejandro Callejas in Xalapa, and Larry Onisto in Toronto.

In 2003, when Susan Burns and I also enticed Eric Frothingham and Steve Goldfinger to think through what should eventually become Global Footprint Network, a new chapter opened up. We started by inviting our heroes to lend their name and wisdom to our new organization, and one of the first people who accepted the invitation was the late Wangari Maathai (1940-2011) who continues to inspire me to this day.

Many amazing people mentored us or served on Global Footprint Network's Policy Advisory Council. They include Alex Hinds, Andrew Simms, Atif Kubursi, Bruno Oberle, Charles McNeill, Chris Hails, Claude Martin, Catherine Parrish, Craig Simmons, Daniel Pauly, David Batker, David Suzuki, Dominique Voynet, Doug Kelbaugh, Edward O. Wilson, Emil Salim, Eric Garcetti, Ernst Ulrich von Weizsäcker, Fabio Feldmann, François Droz, Gianfranco Bologna, Herman Daly, Henry Frechette, Howard Fairbank, James Gustave Speth, Jim Merkel, Jonathan Loh, Jørgen Randers, Juan Alfonso Peña, Julia Marton-Lefèvre, Karen Kraft Sloan, Karl-Henrik Robèrt, Lester Brown, Luc Bas, M.S. Swaminathan, Manfred Max-Neef, Mark Halle, Melita Elmore, Nicky Chambers, Norman Myers, Oscar Arias, Partha Dasgupta, Paul Messerli, Peter Boothroyd, Peter Raven, Peter Wilderer,

Pooran Desai, Rashid Bin Fahad, Rhodri Morgan, Robert Klijn, Roberto Brambilla, Rosalía Arteaga, Sebastian Navarro, Simon Pearson Simon Upton, Simone Bastianoni, Stephen Groff, Terry A'Hearn, Thomas Lovejoy, Uwe Schneidewind, Vicki Robin, Will Steffen, William (Bill) Rees, Wolfgang Sachs, Xavier Houot, and Yoshihiko Wada.

Daring people have served on the Global Footprint Network board. They have thrown themselves in it fully, delivering not just sweat, blood, tears, but also much laughter and many generous donations. These wonderful individuals are Alexa Firmenich, André Hoffmann, Ann Hancock, Bob Doppelt, Cara Pike, Daniel Goldscheider, Elizabeth McNamee, Eric Frothingham, Haroldo Mattos de Lemos, Jamshyd Godrej, John Balbach, Julia Marton-Lefèvre, Keith Tuffley, Kristin Cobble, Louis de Montpellier, Lynda Mansson, Michael Saalfeld, Razan Khalifa Al Mubarak, Rob Lilley, Sandra Browne, Sarosh Kumana, Susan Burns, Terry Vogt, and Tony Long.

The miracle of the execution is the labor of amazing hands. The wonderful people who have worked for Global Footprint Network on our staff include: Adeline Murthy, Alessandro Galli, Amanda Diep, Armando Alves, Anna Oursler, Annabel Hertz, Audrey Kitzes (née Peller), Benjamin Bellman, Bessma Mourad, Birgit Maddox, Bonnie McBain, Brad Ewing, Bree Barbeau, Brooking Gatewood, Carol DiBenedetto, Cylcia Bolibaugh, Chad Monfredo, Chiron Mukherjee, Chris Martiniak, Christopher Nelder, Dana Smirin, Daniel Moran, David Lin, David Moore, David Zimmerman, Debbie Cheng, Denine Giles, Derek Eaton, Dharashree Panda, Diana Deumling, Diane Stark, Drew Lisac, Eli Lazarus, Emily Daniel, Evan Neill, Faith Flanigan, Fatime-Zahra Medouar, Firesenai Sereke, Fouad Hamdan, Francesca Silvestri, Frank Thompson, Gemma Cranston, Geoff Trotter, Giacomo Pascolini, Gina Kiani, Gina DiTommaso, Golnar Zokai, Haley Kingsland, Helena Brykarz, Ian Wymore, Ingrid Heinrich, Jag Alexeyev, James Espinas, Jaime Speed, Jan Schwarz, Jason Ortego, Jennifer Mitchell, Jill Connaway, Jon Martindill, Joy Whalen, Joy Larson, Juan Carlos Morales, Judith Silverstein, Judith Sissener, Juliana Linder, Julie Curry, Justin Kitzes, Kamila Kennedy, Karin Hess, Kath Delaney, Katsunori Iha, Kevin Clark, Krina Huang, Kristin Kane, Kyle Gracey, Kyle Lemle, Kylie Carera, Laetitia Mailhes, Laura Yuenger, Laura Loescher, Laurel Hanscom, Loic Lombard, Loredana Serban, Mahsa Fatemi, Maria Leticia Figueroa, Mark Lancaster, Martin Kärcher, Mariko Meyer, Martin Halle, Mary Thomas, Maxine McMinn, Melanie Hogan, Melissa Fondakowski, Melissa Mazzarella, Meredith Delich (née Stechbart), Michael Borucke, Michael Wang, Michael Murray, Michel Gressot, Michelle Shaffer, Mike Wallace, Mikel Evans, Mimi Torres, Nicole Grunewald, Nicole Freeling, Nina Hausman, Nina Bohlen, Nina Brooks, Olaf Erber, Pablo Muñoz, Pati Poblete, Paul Wermer, Pragyan Bharati, Priyangi Jayasinghe, Rachel Roberts, Rachel Hodara Nelson, Ramesh Narasimhan,

Robert Williams, Ronna Kelly, Ryan Van Lenning, Sandra Browne, Samir Gupta, Sara Friedman, Sarah Rizk, Sarah Drexler, Scott Mattoon, Sebastian Winkler, Selen Altiok, Serena Mancini, Shiva Niazi, Sophia Perez, Steven Goldfinger, Susan Burns, Tarek Saleh, Tatjana Puschkarsky, Tina Batt, Tony Drummond, William (Bill) Coleman, Willy De Backer, and Yves De Soye. Plus hundreds of remarkable interns and volunteers, who brought fresh ideas and enthusiasm to our ventures.

All the work has been made possible by the profound generosity of uncountable individuals, including Caroline Wackernagel, Daniela Schlettwein, Frank and Margrit Balmer, Isabelle Wackernagel, the late visionary and conservation pioneer Luc Hoffmann, Marie-Christine Wackernagel, Peter Seidel, Roland Matter, Ruth and Hans-Edi Moppert, Stephan Schmidheiny, Urs and Barbara Burckhardt, and Urs-Peter Geiger. Foundations that have kindly accompanied us on our path include Asahi Glass Foundation, Atkinson Charitable Foundation, Avina Stiftung, Barr Foundation, Binding-Stiftung, Dr. med Arthur und Estella Hirzel-Callegari Stiftung, Erlenmeyer Stiftung, Flora Family Foundation, Foundation for Global Community, Foundation Harafi, Fundação Calouste Gulbenkian, Furnessville Foundation, Hull Family Foundation, Iverson Family Fund, James Gustave Speth Fund for the Environment, MAVA Foundation, Mental Insight Foundation, Oak Foundation, Richard and Rhoda Goldman Fund, Rockefeller Foundation, Roy A. Hunt Foundation, Skoll Foundation, Stiftung Drittes Millennium, Stiftung Mercator Schweiz, TAUPO Fund, Tellus Mater Foundation, The Dudley Foundation, The Kendeda Fund, The Lawrence Foundation, The Pollux/ProCare Foundation, The Santa Barbara Family Foundation, Town Creek Foundation, Trio Foundation, V. Kann Rasmussen Foundation, Weeden Foundation, and Winslow Foundation. The lasting partnership with WWF, starting in 1996 with their Living Planet Campaign, has also been transformational. The law firm Cooley has been generous in counseling us along the way, on many tricky and exciting initiatives. New partnerships include Schneider Electric, a company that keeps track of how effective they are in helping drive humanity out of overshoot. To them this focus makes basic business sense because radical decarbonization is a service that will be needed ever more.

Many have helped us to think beyond the next corner, including our dear colleagues at York University, with whom we are preparing another great adventure. Thank you Peter Victor, Ravi de Costa, Eric Miller, Martin Bunch, and Alice Hovorka. I am also inspired by my tide-turning friends Lynne and Bill Twist, Ocean and John Robbins, Vicki Robin, Laura Loescher, Neal Rogin, Tracy Apple, Richard Rathbun, Van Jones, Joe Kresse, and Tom Burt. I am grateful too to the many friends of the Balaton Group, and my fellow 99 members and many friends of the Club of Rome.

The warm and wonderful team at New Society Publishers, from Rob West, Sue Custance, Sara Reeves, Greg Green, the kind editor Betsy Nuse all the way to Katharina Rout who provided the initial translation deserve my gratitude. It has been a particular thrill to work again with Phil Testemale, a cherished friend and wonderful colleague, who agreed, in spite of his busy professional life, to spice up the book with a few illustrations. Phil and I started in 1992 to explore the power of explaining our ideas visually. His gift to the cause is tremendous.

And, of course, this all would have been quite soulless without Mathis's entertaining home life with Susan, André, Julia, Alex, and Chester. The same is true for Bert, whose life is tremendously enriched and grounded by Judy and Jil.

This seemingly long list is but a small sample of the people around the world contributing to this work, including you, dear reader of this book. I also want to acknowledge the 3 million footprintcalculator.org visitors just last year, or the 3 billion media impressions Earth Overshoot Day 2018 generated in over 100 countries based on over 2,000 news stories we could track. And that is just the beginning.

For more information about Global Footprint Network, visit our websites:

footprintnetwork.org—offers, as Global Footprint Network's main site, all the background on the Ecological Footprint and its applications you can wish for

data.footprintnetwork.org—provides all key results for the National Footprint Accounts on an open data platform

footprintcalculator.org—allows individuals to estimate their own Footprint and their personal Overshoot Date. Entry point to the #MoveThe-Date map

overshootday.org—hosts Earth Overshoot Day and features solutions to #MoveTheDate

financefootprint.org—highlights the relevance of Ecological Footprint and related results for the finance industry

achtung-schweiz.org/en—applies the Ecological Footprint logic to Switzerland's competitiveness

chinafootprint.org and zujiwangluo.org—provide information about the Ecological Footprint in English and Chinese

OnePlanetAlliance.org—outlines the path forward for independent National Footprint and Biocapacity Accounts

Glossary

Biocapacity: The capacity of ecosystems to regenerate plant matter. Plant matter is essential for life: for instance, plants are at the bottom of every food chain of every animal as well as people. Animals and people compete for Earth's biologically productive space. People use ecosystems to provide food, timber, fiber; to accommodate roads and houses; and to absorb waste material generated by humans. Biocapacity is usually expressed in global hectares. In the National Footprint and Biocapacity Accounts, the biocapacity of an area is calculated by multiplying the actual physical area by the appropriate yield factor and equivalence factors.

Biocapacity Deficit or Reserve: The difference between the biocapacity and Ecological Footprint of a region or country. A *biocapacity deficit* occurs when the Footprint of a population exceeds the biocapacity of their region. Conversely, *biocapacity reserve* exists when the biocapacity of a region exceeds its population's Footprint. If there is a regional or national biocapacity deficit, it means that the region is importing biocapacity through trade, liquidating its regional ecological assets, or using the global commons (for instance by emitting wastes into the global atmosphere or fishing in international waters) to fulfill its needs. In contrast to the national scale, the global biocapacity deficit cannot be compensated through trade, and is therefore equal to overshoot by definition.

Consumption–Land-Use Matrix (CLUM): Starting with data from the National Footprint and Biocapacity Accounts, a Consumption–Land-Use Matrix allocates the six major Footprint area categories to the five basic consumption components. Footprint area categories, shown in column headings, are built-up land, carbon footprint, cropland, grazing land, forests for timber products, and fishing grounds. Consumption, shown in the row headings are food, shelter, mobility, goods, and services. For additional resolution, each consumption component can be disaggregated further. These matrices are often used as a starting point for sub-national (e.g., state, county, city) Footprint assessments. In this case, national data for each cell is scaled up or down depending on the unique consumption patterns in within this sub-national population compared to the national average.

Earth Overshoot Day: The day each calendar year by which humanity's demand has used as much from Earth as Earth's ecosystems can renew in that entire year. In 2019, the day fell on July 29. For more information see overshootday.org.

Ecological Footprint: A measure of how much area of biologically productive land and water an individual, population, or activity requires to produce all the resources it consumes, to accommodate all its infrastructure, and to absorb the waste it generates, using prevailing technology and resource management practices. The Ecological Footprint is usually measured in global hectares. Because trade is global, an individual or country's Footprint includes land or sea from all over the world. Without further specification, Ecological Footprint generally refers to the Ecological Footprint of consumption. Ecological Footprint is often referred to in short form as the Footprint.

Ecological Poverty Trap: Countries, regions, or populations with an biocapacity deficit and less than world income are particularly vulnerable to staying stuck in their poverty as long as overshoot persists, as they do not have enough resources to power their economy, nor enough purchasing power to supplement their resource needs with imports from elsewhere.

Equivalence Factor: A productivity-based scaling factor that compares a specific land type (such as cropland or forest) to a productive area with world-average biological productivity, as represented by the global hectare. For land types (e.g., cropland) with a productivity higher than the average productivity of all biologically productive land and water area on Earth, the equivalence factor is greater than one. Thus, to convert an average hectare of cropland to global hectares, it is multiplied by the cropland equivalence factor, currently estimated at 2.5. Grazing lands, which have lower productivity than cropland, have an estimated equivalence factor of 0.46 according to the newest National Footprint and Biocapacity Accounts (see also Yield Factor). In a given year, equivalence factors are the same for all countries.

Global Hectare (or gha): The unit of measurement for Ecological Footprint and biocapacity accounting. It represents one hectare of biologically productive area of globally average productivity. There are about 12.2 billion hectares of biologically productive area on Earth. This means that each global hectare harbors about 1/12.2th billionth of the Earth's total biocapacity.

Life-Cycle Assessment (LCA): A quantitative method for assessing a product's material inputs and waste flows over its entire lifespan. LCA attempts to quantify what comes in and what goes out of a product from "cradle to grave," including the energy and material associated with materials extraction, product manufacture and assembly, distribution, use and disposal and the environmental emissions that result. LCA applications are governed by the ISO 14040 series of standards (iso.org). When assessing the Ecological Footprint of a product, LCA data is required as input. These amounts then are translated into global hectares to provide a biocapacity interpretation of the material flows associated with the product.

National Footprint and Biocapacity Accounts: The central data set that calculates the Footprint and biocapacity of the world and more than 200 nations from 1961 to the present. This data is based on UN statistics. The results generally come with a three-year lag due to delayed availability of raw statistics. But using partially available data plus estimates, results can be now-casted.

Now-casting: To extrapolate missing data in order to estimate results for today (rather than stopping a couple of years ago). View the most recent data on the Ecological Footprint Explorer open data platform (data.footprintnetwork.org).

Overshoot: Global overshoot occurs when humanity's demand on nature exceeds the biosphere's regenerative capacity or supply. Such overshoot leads to a depletion of Earth's life-supporting natural capital, including the buildup of waste such as ocean acidification from excessive CO_2 or climate change from greenhouse gas accumulation in the atmosphere. At the global level, biocapacity deficit and overshoot are the same, since there is no net import of material resources to the planet. Local overshoot occurs when a local ecosystem is exploited more rapidly than it can renew itself.

Sustainable Development: Sustainable development became a more common policy term through the Brundtland Commission's report to the UN.* These two words conveys the need to overcome a fundamental tension: social desires versus ecological possibilities. "Development" represents people's desires for fulfilling, secure lives. "Sustainable" stands for the recognition that humanity shares this one planet. Therefore, the combination of these two words simply imply that goal for "all to thrive within the means of our one planet."

Tragedy of Open Resource Access: Under certain circumstances, it is possible for individuals to concentrate the benefits, while the costs are socialized. Economists sometimes call this "externalities." This effect is often associated with the inaccurately named article "Tragedy of the Commons." The

Commons, as the original author acknowledged, are a possible solution to the tragedy.

Yield Factor: A factor that accounts for differences in productivity of land within a given land type. In National Footprint and Biocapacity Accounts, each country and each year has yield factors for cropland, grazing land, forest, and fishing grounds. For example, in 2016, German cropland was 1.44 times more productive than world average cropland. The German cropland yield factor of 1.44, multiplied by the cropland equivalence factor of 2.5 converts German cropland hectares into global hectares: one hectare of cropland is equal to 3.6 gha.

* This report proposed a more complex definition of sustainable development that slightly hides the underlying tension between enhancing human well-being and living within Earth's regenerative capacity: "Development that meets the needs of the present without compromising the ability of future generations to meet their own needs." World Commission on Environment and Development, *Our Common Future*, p. 27.

Notes

Prelude

1. See the Glossary for further explanation.

Introduction

1. Peter A. Victor, *Managing Without Growth: Slower by Design, Not Disaster*, 2nd ed. (Cheltenham and Camberley UK: Edward Elgar Publishing, 2019).

2. The top 100 urban areas alone emit 18% of the global CO_2. For details see: Daniel Moran et al., "Carbon Footprints of 13 000 Cities," *Environmental Research Letters* 13, no. 6 (June 19, 2018), accessed February 27, 2019, iopscience .iop.org/article/10.1088/1748-9326/aac72a.

3. Pooran Desai and Paul King, *One Planet Living* (Bristol UK: Alastair Sawday Publishing, 2006).

4. Mitch Hescox with Paul Douglas, *Caring for Creation: The Evangelical's Guide to Climate Change and a Healthy Environment* (Ada MI: Bethany House, 2016).

5. With every new edition of the Ecological Footprint's annual calculations come not only new data (including new historical data) but also some improvements to the accounting methodology. This is why the accounts are recalculated back to 1961 every year. The latest academic description of the accounts is available here: David Lin et al., "Ecological Footprint Accounting for Countries: Updates and Results of the National Footprint Accounts, 2012–2018," *Resources* 7 (2018): 58, mdpi.com/2079-9276/7/3/58.

    The biggest changes typically stem from better data, as the UN also changes its data set historically. For instance, former estimates get replaced by actuals. One significant improvement introduced in the 2016 edition is a more precise estimate of the global forests' capacity to sequester additional carbon. The new estimates showed a lower global average for sequestration than previous estimates had assumed. This change came from better global data on forest productivity, not a change in accounting principles or methodology. The new calculation is documented in an academic paper by Global Footprint Network and University of Siena researchers published in 2016: Serena Mancini et al., "Ecological Footprint: Refining the Carbon Footprint Calculation," *Ecol. Indic.* 61 (2016): 390–403. As a result of this new calculation, carbon Footprints per annual ton of CO_2 emission have increased compared to previous editions.

The latest results of the National Footprint and Biocapacity Accounts are freely available on an open data platform at data.footprintnetwork.org. The 2019 edition of the National Footprint and Biocapacity Accounts is the basis of the Ecological Footprint and biocapacity results presented in this publication.

6. Global Footprint Network, accessed February 27, 2019, www.footprintnetwork .org; "The Ecological Footprint Initiative," York University, accessed February 27, 2019, footprint.info.yorku.ca. The initiative is pushed forward by the One-Planet Alliance. oneplanetalliance.org.

7. Global Footprint Network called this initiative "Ten-in-Ten:" ten countries in ten years.

8. "Country Work," Global Footprint Network, accessed February 27, 2019, footprintnetwork.org/our-work/countries.

9. "Switzerland," Case Studies—Global Footprint Network, January 10, 2017, accessed February 27, 2019, footprintnetwork.org/2017/01/10/switzerland. The results are discussed: Dale Bechtel, "Green initiative will not leave footprint on economy," accessed February 27, 2019, swissinfo.ch/eng/september -25-vote_footprint-of-green-initiative-on-swiss-economy/42465734 or tiny.cc/swiss-green-vote.

Chapter 1

1. Terms like *global hectare* are explained in the Glossary as well as in Chapter 3.

2. To be more precise, the Footprint also includes all biocapacity demands to extract, refine, distribute the fossil fuel.

3. This Footprint also includes all biocapacity demands to extract, refine, distribute the fossil fuel. However, it does not included the portion of carbon dioxide that gets absorbed by the oceans: Mancini, "Ecological Footprint: Refining the Carbon Footprint Calculation."

4. Well-managed cropland and pastureland can also become a carbon sink, as more carbon is brought into the soils. This needs a careful shift in agricultural practice.

5. The Swiss Federal Institute of Technology has been active in pilot tests: "CO_2 capture and storage," Separation Proccesses Laboratory—ETH Zurich, accessed March 13, 2019, spl.ethz.ch/research/co2-capture-and-storage.html.

6. Gretchen Daily, ed., *Nature's Services: Societal Dependence on Natural Ecosystems* (Washington DC: Island Press, 1997).

7. As calculated in the 2019 edition of the National Footprint and Biocapacity Accounts.

8. "The Ecological Footprint Intiative," York University, accessed March 2, 2019, footprint.info.yorku.ca/.

9. "What Is Your Ecological Footprint?" Global Footprint Network, accessed March 2, 2019, footprintcalculator.org/.

10. This calculation is based on the high greenhouse gas equivalent of methane.

11. This estimate is based on the Multi-Regional Input-Output assessment of Global Footprint Network that is built on the National Footprint and Biocapacity Accounts.

Chapter 2

1. The University of Arizona has now taken over the management of Biosphere II. For more details, visit biosphere2.org/.
2. William Rees, "The Regional Capsule Concept: An Heuristic Model for Thinking about Regional Development and Environmental Policy," Concept Note, School of Community and Regional Planning, University of British Columbia, 1986.
3. "City Limits," City of London, 2002, accessed January 29, 2019, citylimitslondon.com/.
4. This result comes from a 2003 study. The current Footprint of London will certainly have changed since.
5. William Stanley Jevons, *The Coal Question* (London and Cambridge: MacMillan, 1865), 306.
6. "Footprint Standards," Global Footprint Network, accessed March 3, 2019, footprintnetwork.org/resources/data/footprint-standards/.
7. The idea for such a "meditation" comes from Richard Heinberg, *The Party's Over: Oil, War, and the Fate of Industrial Societies* (Gabriola Island BC: New Society, 2005), 186.
8. Heinberg, *The Party's Over*, 195.
9. The following passage is based on Herbert Girardet, "Die Schaffung lebenswerter und nachhaltiger Städte" in Herbert Girardet, ed., *Zukunft ist möglich: Wege aus dem Klima-Chaos* (Hamburg: Europäische-Verlagsanstalt, 2007), 182ff. A similar English version is Herbert Girardet, *Cities People Planet: Urban Development and Climate Change*, 2nd ed. (New Jersey: Wiley, 2008).
10. "Urbanization of the United States from 1945," Demographia, accessed January 29, 2019, demographia.com/db-1945uza.htm.
11. "Major Agglomerations of the World," City Population, accessed March 14, 2019, citypopulation.de/world/Agglomerations.html.
12. "2018 Revision of World Urbanization Prospects," UN Department of Economic and Social Affairs, accessed March 13, 2019, un.org/development/desa/publications/2018-revision-of-world-urbanization-prospects.html.
13. "Major Agglomerations of the World," City Population.
14. See "Demographics" population tables (based on the US Census) in Wikipedia entry *Manhattan*, accessed March 13, 2019, en.wikipedia.org/wiki/Manhattan#Demographics.
15. Ernst-Ulrich von Weizsäcker, Charlie Hargroves et al., *Factor Five: Transforming the Global Economy Through 80% Improvements in Resource Productivity* (London: Earthscan, 2011).

Chapter 3

1. Vaclav Smil, *The Earth's Biosphere: Evolution, Dynamics, and Change* (Cambridge MA: MIT, 2003), 19.
2. For a definition of *primary productivity*, see Eugene P. Odum's classic *Fundamentals of Ecology*.
3. Different systems exist to estimate how much of the biomass production of ecosystems is used by humans. According to the 2007 study cited below,

humanity claimed ¼ of the planet's entire terrestrial net primary productivity for itself. That is a significant amount for just one species, and the trend is upward. We also need to consider that ecological calamities may occur before 100% of the capacity is appropriated. The Footprint helps because it compares the actual human consumption of biocapacity with the available biocapacity that can potentially be consumed on a renewable basis. Aligned with E. O. Wilson, we assert that the goal should not be to use 100% of Earth's biocapacity as wild species need space, too. See Helmut K. Haberl, K. Heinz Erb, et al., "Quantifying and Mapping the Human Appropriation of Net Primary Production in Earth's Terrestrial Ecosystems," *Proceedings of the National Academy of Sciences* 104, no. 31 (July 2007): 12942–12947, doi:10.1073/pnas.0704243104.

   The net primary productivity assessments help to estimate how intensely ecosystems are being used. The NPP related methods, however, produce less precise results when used to compare ecological production with human harvesting.

4. How humans use and benefit from ecological goods and services has changed throughout history. The ongoing work toward optimizing Footprint standards takes that fact into account.

5. UN Food and Agriculture Organization provides a core repository of large data sets on land use productivity and other critical data points: "Land Use," UNFAO, cited March 16, 2019, fao.org/faostat/en/#data/RL/metadata.

6. Skillfully managed cropland and pasture grassland can also become a carbon sink, and this can be accelerated as more carbon is brought into the soils. This needs a careful shift in agricultural practice, as discussed in "Challenges and opportunities for carbon sequestration in grassland systems," UN FAO, 2009, accessed March 13, 2019, fao.org/fileadmin/templates/agphome/documents/climate/AGPC_grassland_webversion_19.pdf.

   This carbon sequestration service of grasslands could also be included in Ecological Footprint accounts, but is not yet. See for instance: Kat Kerlin, "Grasslands More Reliable Carbon Sink Than Trees," UC Davis, July 9, 2018, accessed March 13, 2019, ucdavis.edu/news/grasslands-more-reliable-carbon-sink-trees.

   Both forests and grasslands need to be managed carefully to make sure they truly are lasting carbon sinks: M. B. Jones and A. Donnelly, "Carbon Sequestration in Temperate Grassland Ecosystems and the Influence of Management, Climate and Elevated CO_2" *New Phytologist*, 164: 423–439. doi:10.1111/j.1469-8137.2004.01201.x.

7. IPCC, *Special Report: Global Warming of 1.5°C—Summary for Policymakers*, paragraph D.1.1, accessed March 15, 2019, ipcc.ch/sr15/.

8. Technical details can be found in the various documents available at footprintnetwork.org/resources/data. This web page includes a preprint of a method paper (Michael Borucke et al., "Accounting for Demand and Supply of the Biosphere's Regenerative Capacity: The National Footprint Accounts' Underlying Methodology and Framework," accessed March 3, 2019, footprint network.org/content/images/NFA%20Method%20Paper%202011%20

Submitted%20for%20Publication.pdf) as well as a detailed handbook that describes the national calculation template underpinning the National Footprint and Biocapacity Accounts.

9. Apart from a few exceptions. For example, there are ecosystems in the deep sea that are sustained by geothermal energy.

10. Robert E. Blankenship, "Future Perspectives in Plant Biology—Early Evolution of Photosynthesis," *Plant Physiology* 154 (October 2010): 434–438, plantphysiol.org/content/plantphysiol/154/2/434.full.pdf, doi: https://doi .org/10.1104/pp.110.161687.

11. Yuval Noah Harari, *Sapiens: A Brief History of Humankind* (New York NY: Harper, 2015), 65.

12. Yinon M. Bar-On, Rob Phillips, and Ron Milo, "The Biomass Distribution on Earth," *Proceedings of the National Academy of Sciences* 115, no. 25 (June 19, 2018): 6506–6511, doi.org/10.1073/pnas.1711842115.

13. UN Environment Program, World Conservation Monitoring Program, and International Union for Conservation of Nature, *Protected Planet Report 2016*, accessed January 31, 2019, wdpa.s3.amazonaws.com/Protected_Planet _Reports/2445%20Global%20Protected%20Planet%202016_WEB.pdf.
    Well-managed cropland and pastureland can also become a carbon sink, as more carbon is brought into the soils. This needs a careful shift in agricultural practice.

14. Fred Pearce, *When the Rivers Run Dry: What Happens When Our Water Runs Out?* (Boston MA: Beacon, 2007).

15. Peter Kareiva et al., "Domesticated Nature: Shaping Landscapes and Ecosystems for Human Welfare," *Science* 316, no. 5833 (June 29, 2007): 1866–1869, science.sciencemag.org/content/316/5833/1866.full.

16. H. Schandl et al., *Global Material Flows and Resource Productivity: An Assessment Study of the UNEP International Resource Panel*. United Nations Environment Programme, 2016, accessed January 31, 2019, esourcepanel.org/reports /global-material-flows-and-resource-productivity-database-link. Also see Stefan Bringezu and Raimund Bleischwitz, eds., *Sustainable Resource Management: Global Trends, Vision and Policies* (London: Routledge, 2009).

17. The following historical sketch mostly follows Rolf Peter Sieferle et al., *Das Ende der Fläche: Zum gesellschaftlichen Stoffwechsel der Industrialisierung* (Vienna: Böhlau, 2006). See also Jill Jäger, *Our Planet: How Much More Can Earth Take?* (London: Haus, 2009).

18. M. Dittrich et al., 2012. "Green Economies Around the World?: Implications of Resource Use for Development and the Environment," SERI, Vienna, 2012, accessed March 13, 2019, boell.de/en/ecology/publications-green-economies -around-the-world-15194.html.

19. Yinon M. Bar-On, Rob Phillips, and Ron Milo, "The Biomass Distribution on Earth," *PNAS* 115, no. 25 (June 19, 2018): 6506–6511, doi.org/10.1073/pnas .1711842115.

20. Schandl, *Global Material Flows and Resource Productivity*.

21. Daniel Yergin, *The Prize: The Epic Quest for Oil, Money and Power* (New York: Simon & Schuster, 2003), 494 ff and 541 ff.

22. Eillie Anzilotti, "Food Waste Is Going to Take Over the Fashion Industry," *Fast Company*, June 15, 2018 accessed February 2, 2019, fastcompany.com /40584274/food-waste-is-going-to-take-over-the-fashion-industry.

23. NOAA, the US National Oceanic & Atmospheric Administration, provides constant updates: "Trends in Atmospheric Carbon Dioxide," accessed March 13, 2019, esrl.noaa.gov/gmd/ccgg/trends/.

24. This term was coined by Friedrich Schmidt-Bleek, former vice-president of the Wuppertal Institute and author of *The Earth: Natural Resources and Human Intervention* (London: Haus, 2009).

25. Edward Osborne Wilson is an American biologist and entomologist who specializes in evolutionary theory and sociobiology. He is an authority on ants and is seen as the intellectual "Father of Biodiversity." See E. O. Wilson, *The Diversity of Life* (Cambridge: Harvard, 1992). In 2003, he wrote *The Future of Life* (New York: Knopf), which included a footnote on the idea of dedicating half of planet Earth to biodiversity. *Half-Earth: Our Planet's Fight for Life* (New York: Liveright, 2016) turned into the half-earthproject.org.

Chapter 4

1. Lester B. Brown, *Outgrowing the Earth: The Food Security Challenge in an Age of Falling Water Tables and Rising Temperatures* (New York: W.W. Norton, 2005).

2. Jil, Bert Beyers' daughter, was born in 1997, and André, Mathis Wackernagel's son, in 2001.

3. Franz Josef Radermacher used his training in mathematics and intercon-nected systems to study the extraordinarily successful development of the superorganism humanity and how that development has been driven by communication, interaction, and technological innovation. From that same perspective, he also analyzed the historic transition humanity will face in the 21st century. This much is certain: The growth pattern in humanity's devel-opment that has existed for millennia will break down, and another pattern will replace it. At the same time, the superorganism humanity runs the risk of overburdening its biotope, Earth, beyond its natural ecosystem sustainability. See Franz Joseph Radermacher and Bert Beyers, *Welt mit Zukunft: Überleben im 21. Jahrhundert* (Hamburg: Murmann, 2007). One of Radermacher's English publications is *Global Marshall Plan—A Planetary Contract: For a Worldwide Eco-Social Market Economy* (Hamburg: Global Marshall Plan Initiative, 2004).

4. Core Writing Team, R. K. Pachauri and L. A. Meyer, eds., *Climate Change 2014: Synthesis Report. Contribution of Working Groups I, II and III to the Fifth Assessment Report of the Intergovernmental Panel on Climate Change* (Geneva, Switzerland: IPCC 2014).

5. James H. Butler and Stephen A. Montzka, *The NOAA Annual Greenhouse Gas Index (AGGI)* (Boulder CO: NOAA Earth System Research Laboratory, Spring 2018), accessed January 2, 2019, esrl.noaa.gov/gmd/aggi/aggi.html.

6. Simon Upton, personal communication with Mathis.

7. J. Kitzes et al., "A Research Agenda for Improving National Ecological Foot-print Accounts," *Ecological Economics* 68, no. 7 (2009): 1991–2007.

8. For technical details, see "Data and Methodology," Global Footprint Network, accessed March 4, 2019, footprintnetwork.org/resources/data/.

9. Emily Matthews et al., *The Weight of Nations: Material Outflows from Industrial Economics* (Washington DC: World Resource Institute, 2000). See also wri.org.

10. Part II addresses who has access to what resources, who will be the winners and who the losers, as well as how geopolitical power might shift in the future.

11. Mathis Wackernagel et al., "Defying the Footprint Oracle: Implications of Country Resource Trends," *MDPI-Sustainability* 2019, 11, 2164. (accessed April 22, 2019) mdpi.com/2071-1050/11/7/2164.

12. The numbers are derived from data sets by the United Nations' Food and Agricultural Organization (FAO) and apply to the year 2016. Land use statistics of FAO are available at fao.org/faostat/en/#data/EL.

13. Chris Chisholm, "The Food Insecurity of North Korea," National Public Radio, June 19, 2018, accessed March 12, 2019, npr.org/sections/goatsandsoda/2018/06/19/620484758/the-food-insecurity-of-north-korea.

14. Peter H. Gleick, "Water, Drought, Climate Change, and Conflict in Syria," *Weather Clim Soc.*, 6, no.3 (July 2014): 331–340, accessed March 12, 2019, doi.org/10.1175/WCAS-D-13-00059.1.

15. Clive Ponting, *A New Green History of the World: The Environment and the Collapse of Great Civilisations* (London: Penguin, 2007), 71ff.

16. Donella Meadows, Jorgen Randers, and Dennis Meadows, *Limits to Growth: The 30-Year Update* (White River Junction VT: Chelsea Green Publishing, 2004).

17. Note that numerous pressures, such as deforestation or soil loss, are not adequately described by the UN data sets used for the National Footprint and Biocapacity Accounts, leading to undercounting.

18. Following the American economist Herman Daly, we can describe the historical phase before we reached global overshoots as "empty world." During that period, the supply of both area and resources was apparently not a growth-limiting factor. Annual flows of resources and energy were more than sufficient to satisfy human demands. Since the mid-1970s we have been in the phase of a "full world," with area and resources becoming more constrained. See Herman E. Daly, "Towards Some Operational Principles of Sustainable Development," *Ecological Economics* 2, no. 1: 1–6. See also Alessandro Galli, "Assessing the Role of the Ecological Footprint as Sustainability Indicator" (Dissertation, University of Siena, 2007).

19. Earth Overshoot Day. overshootday.org.

20. One day in a year is less than ⅓ of a percent. Since data behind the National Footprint and Biocapacity Accounts lack that level of precision, the day represents our best estimate. It does not reflect the level of precision of the accounts. The actual day may fall within a couple of weeks. However, comparisons between years are more accurate.

21. Josef H. Reichholf, *Ende der Artenvielfalt? Gefährdung und Vernichtung von Biodiversität* (Frankfurt a. M.: Fischer-Taschenbuch-Verlag, 2008), 130.

22. The connections between the Footprint and the Living Planet Index were initially promoted by WWF's Living Planet Reports, starting in the year 2000. The Living Planet Index now tracks 4,270 species and 21,252 populations of mammals, birds, reptiles, amphibians, and fish; this index was developed initially by WWF. It is now hosted by the Zoological Society of London:

livingplanetindex.org. The regularly updated Living Planet Report can be found here: footprintnetwork.org/living-planet-report.

23. The 10% goal dates back to the World Parks Congress in Bali, 1982. See also "Strategic Plan for Biodiversity 2011–2020, including Aichi Biodiversity Targets," Convention on Biological Diversity, accessed February 5, 2019, cbd.int/sp/.

24. Reichholf, *Ende der Artenvielfalt*, 142.

25. Reichholf, *Ende der Artenvielfalt*, 163ff.

26. In reality, international development investments into family planning through bilateral programs still make up less than 1% of the overall budgets: "Investing in Women and Girls," OECD Development Co-operation website, accessed March 12, 20019, oecd.org/dac/gender-development/investingin womenandgirls.htm.

27. Jim Yong Kim, "To Build a Brighter Future, Invest in Women and Girls," *Voices: Perspectives on Development*, March 8, 2018, accessed March 12, 2019, blogs.worldbank.org/voices/build-brighter-future-invest-women-and-girls.

28. von Weizsäcker, *Factor Five*; Wuppertal Institut für Klima, Umwelt, Energie, *Zukunftsfähiges Deutschland in einer globalisierten Welt: Ein Anstoß zur gesellschaftlichen Debatte* (Frankfurt a. M.: Taschen, 2008); Wolfgang Sachs, *Nach uns die Zukunft: Der globale Konflikt um Gerechtigkeit und Ökologie* (Frankfurt a. M: Taschen, 2003).

Chapter 5

1. In 1961 (the earliest data point for the National Footprint and Biocapacity Accounts), the carbon Footprint made up 44% of humanity's Footprint. In 2016, it had grown to 60%. In Somalia, Niger, Lao, or Myanmar, the carbon Footprint makes up 3%, 5%, 9%, 10% respectively of the total Footprint. In Japan, Switzerland, or the United States, it makes up 75%, 74%, and 70% respectively. The Footprint share for food is 23% in the United States, 29% in Switzerland, and 35% in Germany. In low-income countries, the Footprint share for food is much higher: 74% in Morocco and 80% in Bolivia.

2. World Commission on Environment and Development. *Our Common Future* (Oxford UK: Oxford, 1987). This report fed into the 1992 UNCED conference (United Nations Conference on Environment and Development) in Rio de Janeiro which became the hitherto largest gathering of heads of state. The unstoppable Canadian chair, Maurice Strong, set the agenda on a number of environmental issues from climate to biodiversity, and also more forcefully brought NGOs, businesses, and indigenous communities into the global conversation.

3. Kate Raworth, *Doughnut Economics: Seven Ways to Think Like a 21st-Century Economist* (London: Random House Business, 2017). You can find out more about the doughnut here: kateraworth.com/doughnut/.

4. "Human Development Reports," United Nations Development Programme, accessed February 11, 2019, hdr.undp.org.

5. This chart was also published in UNDP's 2013 Human Development Report (their "Figure 1.7").

6. "Table 2. Human Development Index Trends, 1990–2017" of the 2018 Statistical Update, hdr.undp.org/en/composite/trends.

7. Global Footprint Network's data platform allows tracking countries over time: see "Ecological Footprint Explorer," accessed February 11, 2019, data.footprint network.org.

8. Mathis Wackernagel, Laurel Hanscom, and David Lin, "Making the Sustainable Development Goals Consistent with Sustainability," *Front. Energy Res.* (July 11, 2017), doi.org/10.3389/fenrg.2017.00018.

9. "How Much Nature Do We Have? How Much Do We Use?" Mathis Wackernagel, TedX San Francisco, December 22, 2015, accessed February 11, 2019, youtube.com/watch?v=3M29BY86bP4.

10. In Europe, resource productivity grows by about 2% annually. The result is a relative decoupling (less use of material per value-added unit) but not an absolute decoupling (less material used overall).

11. WWF and Global Footprint Network, "EU Overshoot Day—Living Beyond Nature's Limits: 10 May 2019" WWF-EPO, Brussels, 2019.

12. Reichholf, *Ende der Artenvielfalt*, 132. See also Smil, *The Earth's Biosphere*.

13. The passage about Manila's bat people is based on a report by Wolfgang Uchatius in *Die Zeit* (December 16, 2004).

14. For a more in-depth discussion of China's Footprint, see Chapter 11.

15. Gerhard Lichtenthäler, "Water Conflict and Cooperation in Yemen," *Middle East Report* 254 (Spring 2010), accessed March 16, 2019, merip.org/2010/03/water-conflict-and-cooperation-in-yemen/.

16. "When the Last Tree Is Cut Down…," *Quote Investigator*, accessed February 12, 2019, quoteinvestigator.com/2011/10/20/last-tree-cut/.

Chapter 6

1. Ponting, *A New Green History of the World*, 144ff.

2. Ponting, *A New Green History of the World*, 151ff.

3. Garret Hardin, "The Tragedy of the Commons," *Science* 162, issue 3859 (December 13, 1968): 1243–1248.

4. The solution Hardin offers to overcome his identified tragedy is, in his own words, "mutual coercion, mutually agreed upon." These five words are also the most succinct and poignant definition of a commons. In the literature, though, it is not recognized by many (possibly because of the inaccurate title he may have chosen to provoke the readers) that Hardin proposed commons as a solution.

5. The 2009 Nobel Prize in Economics was awarded to the American Elinor Ostrom, who specialized in problems of common-pool resources and their solutions. For her work, she interviewed alpine farmers and fishers in numerous countries. Unfortunately, she passed away in 2012.

6. This is calculated from the 2019 National Footprint and Biocapacity Accounts. It is simply the sum of all the global overshoot percentages since the first year of overshoot.

7. Jared Diamond, *Collapse: How Societies Choose to Fail or Succeed*, rev. ed. (New York: Penguin, 2011), 79ff.

Chapter 7

1. BakBasel Economics Institute and Global Footprint Network, *The Significance of Global Resource Availability to Swiss Competitiveness* (Basel, Switzerland: BakBasel, 2014), accessed February 13, 2019. Also consult the WEF report 2018 which included the Footprint. tiny.cc/83p86y

2. FOEN (Swiss Federal Office for the Environment), *Umwelt-Fussabdrücke der Schweiz (Environmental Footprints of Switzerland)* (Bern: Bundesamt für Umwelt BAFU, 2018), 87, accessed on May 24, 2019, on bafu.admin.ch or directly at tiny.cc/BAFUfootprint2018

3. Examples are available at "Country Work," Global Footprint Network.

4. "Ecological Footprint Explorer," Global Footprint Network.

5. The data stem from the 2010 Global Burden of Disease Study: "Global Burden of Disease," *The Lancet*, accessed February 13, 2019, thelancet.com/gbd.

6. All Ecological Footprint and biocapacity data in this book are taken from the 2019 edition of the National Footprint and Biocapacity Accounts.

7. The report with the Guizhou government and the corresponding results are available at chinafootprint.org, accessed February 13, 2019.

8. The following description is based on Molly O'Meara Sheehan, ed., *State of the World 2007: Our Urban Future* (New York: W.W. Norton, 2007), 64 ff.

9. Richard Heinberg, *Peak Everything: Waking up to the Century of Declines* (Gabriola Island: New Society, 2007).

10. Michael Jacobs, *The Green Economy: Environment, Sustainable Development and the Politics of the Future* (Concord MA: Pluto Press, 1991), viii.

11. Our reports are available at Finance for Change, accessed February 14, 2019, footprintfinance.org.

12. Our World in Data, accessed February 22, 2019, ourworldindata.org/grapher /world-gdp-over-the-last-two-millennia?time=1..2015.

13. Esteban Ortiz-Ospina, Diana Beltekian, and Max Roser, "Trade and Globalization," Our World in Data, accessed February 22, 2019, ourworldindata.org /trade-and-globalization.

Chapter 8

1. According to the highlights of the *Carbon Budget 2018* report issued by the Global Carbon Project, "Emissions in 2017 were 9.9±0.5 (36.2 $GtCO_2$) with a share of coal (40%), oil (35%), gas (20%), cement (4%), and flaring (1%). Global emissions in 2018 are projected to increase by more than 2% (+1.8% to +3.7%) after three years of almost no growth, reaching 10.1±0.5 GtC (37.1 $GtCO_2$), a new record high." Accessed February 19, 2019, globalcarbon project.org/carbonbudget/index.htm. An overview is also available by C. Figueres et al., "Emissions Are Still Rising: Ramp up the Cuts," *Nature* (December 5, 2018), doi.org/10.1038/d41586-018-07585-6.

2. "CO_2 Emissions 2017," Global Carbon Atlas, accessed February 14, 2019, globalcarbonatlas.org/en/CO2-emissions.

3. The abbreviation "ppm" stands for parts per million. A gas concentration of 1 ppm means this gas has a prevalence of one in a million molecules.

4. Butler and Montzka, *The NOAA Annual Greenhouse Gas Index*.

5. The data stem from the 2010 Global Burden of Disease Study: "Global Burden

of Disease," *The Lancet*, accessed February 13, 2019, thelancet.com/gbd and Jing Huang, Xiaochuan Pan, Xinbiao Guo, Guoxing Li, 2018, Health Impact of China's Air Pollution Prevention and Control Action Plan: An Analysis of National Air Quality Monitoring and Mortality Data, The Lancet Planetary Health, Elsevier, thelancet.com/journals/lanplh/article/PIIS2542-5196(18)30141-4/fulltext.

6. James Hansen et al., "Target Atmospheric CO_2: Where Should Humanity Aim?" accessed February 15, 2019, arxiv.org/ftp/arxiv/papers/0804/0804.1126.pdf.

7. John Marty Anderies et al., "The Topology of Non-Linear Global Carbon Dynamics: From Tipping Points to Planetary Boundaries," *Environ. Res. Lett.* 8, no. 4 (2013) 044048 (13pp), doi:10.1088/1748-9326/8/4/044048.

8. Andrew C. Baker, Peter W. Glynn, Bernhard Riegl, "Climate Change and Coral Reef Bleaching: An Ecological Assessment of Long-Term Impacts, Recovery Trends and Future Outlook," *Estuarine, Coastal and Shelf Science* 80, no. 4 (December 10, 2008), 435–471, or Terry P. Hughes et al., "Spatial and Temporal Patterns of Mass Bleaching of Corals in the Anthropocene," *Science* 359, no. 6371 (January 5, 2018), 80–83, doi: 10.1126/science.aan8048.

9. "Eliminating our carbon Footprint is possible, and filled with opportunities, if we follow the right strategy," Earth Overshoot Day, accessed February 15, 2019, overshootday.org/energy-retrofit.

10. Jevons, *The Coal Question*, 123–124: "It is wholly a confusion of ideas to suppose that the economical use of fuel is equivalent to a diminished consumption. The very contrary is the truth.[...] It is the very economy of its use which leads to its extensive consumption."

Chapter 9

1. A more thorough introduction to Ecological Footprint principles and applications is available in two chapters by M. Wackernagel et al., "Chapter 16: Ecological Footprint Accounts: Principles," and "Chapter 33: Ecological Footprint Accounts: Criticisms and Applications," both in *Routledge Handbook of Sustainability Indictors*, ed. Simon Bell and Stephen Morse (London: Routledge, 2018), 244–264 and 521–539.

2. The basics of these assessments are explained at "Footprint Standards," Global Footprint Network.

3. See ICLEI—Local Government for Sustainability, iclei.org, or C40 Cities, c40.org, both accessed February 18, 2019.

4. The moratorium may only have held for a year. But interestingly it was proposed not by the environment department but the financial planners who started to recognize that economically it was an unfavorable proposition to add more low-density housing, given infrastructure costs, rather than doing infills which allow to use existing infrastructure more effectively. See "Calgary," Global Footprint Network, accessed February 18, 2019, footprintnetwork.org/2015/04/10/calgary/.

5. "C40 Cities" (initiated by New York's former Mayor Michael Bloomberg), c40.org. Ecocity Builders empowers grassroot movements to build sustainable cities and has many resources available: "Ecocity Builders," accessed February 18, 2019, ecocitybuilders.org.

6. David Thorpe, *'One Planet' Cities: Sustaining Humanity within Planetary Limits* (London: Taylor & Francis, 2019).

7. Regarding National Footprint and Biocapacity Accounts, see Chapter 3. More about Footprint methodology can be found at footprintstandards.org.

8. Updates on this new effort can be found under the umbrella of the new initiative "One-Planet Alliance" and "The Ecological Footprint Initiative," York University.

9. The Global Trade Analysis Project (GTAP), maintained by a global network of researchers and policymakers, aims to conduct quantitative analysis of international policy issues. GTAP is housed by the Center for Global Trade Analysis in Purdue University's Department of Agricultural Economics: "Global Trade Analysis Project," accessed February 19, 2019, gtap.agecon.purdue.edu /default.asp.

10. See Glossary.

11. The scores are explained on the Global Footprint Network website or on the freely downloadable data package with all country results available from data.footprintnetwork.org.

12. For more details, see footprintstandards.org.

13. See Part I: "Footprint—The Tool"

14. See "Country Work," Global Footprint Network.

15. Life-Cycle Assessments (LCA) are governed by ISO14040 Standards: "ISO 14040:2006—Environmental management—Life cycle assessment—Principles and framework," International Organization for Standardization, accessed February 19, 2019, iso.org/standard/37456.html.

16. You can download the tool from the website atmobitool.ch/de/tools/mobi tool-faktoren-25.html. It is in German, but with online translation, it's easy to comprehend. The spreadsheet allows you to play with all kinds of options and evaluate the resource demand of various mobility options. Mobitool was initially produced for the Swiss Railways.

17. These numbers were estimated by Jorgen Vos for Global Footprint Network and are based on Volkswagen Jetta and Toyota Prius, based on the EPA 2006 fuel mileage rating system.

18. See "The GPT Group," accessed February 19, 2019, gpt.com.au. The calculator can be accessed at: "Plan8iQ Software," accessed February 19, 2019, gpttreads lightly.com.au/.

19. "GPT," The Footprint Company, accessed February 19, 2019, footprintcompany .com/casestudy/gpt/.

20. Discussion of this #MoveTheDate calculation is available on the Earth Overshoot Day website at overshootday.org/energy-retrofit/ or for more details at schneider-electric.app.box.com/s/1pjh4gdlabgo7m8bfjfhx5jmvsegykxm or tiny.cc/schneider-21days (all were accessed on February 26, 2019).

21. "Eliminating Our Carbon Footprint Is Possible, and Filled with Opportunities, If We Follow the Right Strategy," Global Footprint Network, accessed February 28, 2019, overshootday.org/energy-retrofit. This article is based on case study data from Schneider Electric.

22. Marcus Craig, "To Compete Globally, Dallas County Looks to Efficiency to Fund Long-Term Infrastructure Plans," Schneider Electric Blog, accessed

February 28, 2019, blog.schneider-electric.com/sustainability/2018/05/15
/to-compete-globally-dallas-county-looks-to-efficiency-to-fund-long-term
-infrastructure-plans. A fuller publication, with reference to two projects
in counties (Dallas and Elmore counties) is available in the ebook *A Tale of
Two Communities*, accessed February 28, 2019, hub.resourceadvisor.com
/performance-contracting/a-tale-of-two-communities-ebook.

23. International Development Enterprises—India, accessed February 20, 2019,
    ide-india.org.
24. GramVikas, accessed February 20, 2019, gramvikas.org.

Chapter 10

1. Zero (fossil) Energy Development (ZED), accessed February 20, 2019,
   zedfactory.com.
2. Personal conversation in interview with Bert Beyers, 2011.
3. Nicky Chambers and Craig Simmons founded Best Foot Forward in 1997, the
   first consultancy in Europe to focus on ecological footprint accounting. This
   consultancy is now part of Anthesis Group, anthesisgroup.com.
4. Peabody is a large housing association in London with a long history of
   advocacy for social housing: Peabody housing association London, accessed
   February 20, 2019, peabody.org.uk/home.
5. Bioregional is an organization promoting practical solutions for sustainability: "Bioregional—Championing a Better Way to Live," accessed February 20,
   2019, bioregional.com/.
6. These are ZEDliving's four principles: Make carbon history. Design out fossil
   fuels. Reduce demand—run on native renewables. Enable a high quality of
   life on a LOW Footprint. Bill Dunster, Craig Simmons, and Bobby Gilbert, *The
   ZEDbook: Solutions for a Shrinking World* (Surrey: Taylor and Francis, 2008), 61.
7. "One Planet Living," Bioregional, accessed February 20, 2019, bioregional
   .com/one-planet-living.
8. "Masdar City," Masdar, accessed February 20, 2019, masdar.ae/en/masdar-city.
9. Check out the buildings at masdar.ae/en/masdar-city.
10. For more about Peter's contributions, check out "Peter Seidel—Environmen-
    talist," accessed February 22, 2019, peterseidelbooks.com.
11. Founded in 1919, Bauhaus assembled the most progressive and impactful
    artists of Germany and beyond. As practitioners and teachers, they revolu-
    tionized arts and crafts from architecture all the way to typography. Their
    internationalist and leftist perspective clashed with the Nazi Party. As the
    Nazi Party came into power, operating Bauhaus became untenable. In 1933,
    Mies van der Rohe, its last director, decided to close the school and emigrated
    to the US. He became a leading academic and architect in Chicago. Together
    with Alvar Aalto, Le Corbusier, Walter Gropius, and Frank Lloyd Wright, he is
    credited as the originator of modernist architecture.
12. Harrison Brown, James Bonner, and John Weir, *The Next Hundred Years: Man's
    Natural and Technological Resources; A Discussion Prepared for Leaders of Ameri-
    can Industry* (New York, Viking, 1957).
13. Harrison Brown, *The Challenge of Man's Future: An Inquiry Concerning the Con-
    dition of Man during the Years that Lie Ahead* (New York: Viking, 1954).

14. Peter Seidel's urban designs on "New Communities" are described on his web-site at peterseidelbooks.com/?page_id=8. His professional insights on city design are described in his 1998 article "The Cost of Wealthy Modern Cities" atpeterseidelbooks.com/?page_id=168.

15. Peter Seidel, *Invisible Walls: Why We Ignore the Damage We Inflict on the Planet—And Ourselves* (Amherst NY: Prometheus Books, 1998).

Chapter 11

1. The 13th Five-Year Plan on National Economic and Social Development of the People's Republic of China, March 17, 2016, accessed February 23, 2019, gov. cn/xinwen/2016-03/17/content_5054992.htm. Translation in English is available here: en.ndrc.gov.cn/newsrelease/201612/P020161207645765233498. pdf. While having a strong focus on economic development, the plan also has an impressive number of references to ecological and resource concepts. In the 219 pages of the English translation, the following words appear many times: nature 7; environment 238, resource 217 ecolog- 128, energy 150, food 26, water 185, carbon 12, land 215, climate change 15.

2. "The Guizhou Footprint Report: Metrics for an Ecological Civilization," Global Footprint Network, accessed February 23, 2019, footprintnetwork.org/content /documents/2016_Guizhou_Report_English.pdf.

3. An easy way to compare country performances is available through Global Footprint Network's freely available data package: "Free Public Data Set," Global Footprint Network, accessed February 23, 2019, footprintnetwork.org /licenses/public-data-package-free-2018.

4. Steve Yim Hung-lam, an assistant professor in the Geography and Resources Management Department working at the Chinese University of Hong Kong, is the lead investigator on a published report in the scientific journal *Environmental Research Letters*: Y. Gu et al., "Impacts of Sectoral Emissions in China and the Implications: Air Quality, Public Health, Crop Production, and Economic Costs," *Environmental Research Letters* 13, no. 8 (July 27, 2018), doi: 10.1088/1748-9326/aad138.

5. Charlie Parton, "China's Looming Water Crisis," Chinadialogue, London (May 9, 2018), accessed February 23, 2019, chinadialogue.net/reports/10608-China -s-looming-water-crisis/en.

Chapter 12

1. Wackernagel, "Defying the Footprint Oracle."

2. Stephen Smith, *La ruée vers l'europe: la jeune Afrique en route pour le Vieux Continent* (Paris: Éditions Grasset, 2018).

3. "Africa Ecological Footprint Report 2012," World Wildlife Fund, May 31, 2012, accessed February 23, 2019, wwf.org.za/?6242/aefreportdoc.

4. UN Department of Economic and Social Affairs—Population Division, "World Population Prospects: The 2017 Revision, Key Findings and Advance Tables," Working Paper No. ESA/P/WP/248, accessed February 23, 2019, population .un.org/wpp/Publications/Files/WPP2017_KeyFindings.pdf.

5. Beth Polidoro et al., "Red List of Marine Bony Fishes of the Eastern Central At-

lantic." Gland, Switzerland: IUCN, accessed March 10, 2019, portals.iucn.org /library/node/46290 or http://dx.doi.org/10.2305/IUCN.CH.2016.04.en. "The study highlights the severely limited capacity for fisheries surveillance and enforcement in the region, leading to illegal fishing and overfishing that imperils national and regional management efforts. In many countries, illegal catches represent over 40% of the reported legal catch," the report states.

6. WWF, "Tanzania's Disappearing Timber Revenue," May 25, 2007 accessed March 10, 2019, gftn.panda.org/newsroom/?103600/Tanzanias -disappearing-timber-revenue. The full report published by TRAFFIC in 2007 can be accessed here: d2ouvy59p0dg6k.cloudfront.net/downloads/traffic _tanzania_loggingreport.pdf.

7. See for instance, illegal-logging.info/regions/tanzania. *The Guardian* reported in January 2015 in its article "Tanzania: Illegal Logging Threatens Tree Species with Extinction" that "70% of wood harvested in forests is unaccounted for," *The Guardian*, January 14, 2015, accessed March 10, 2019, theguardian.com /global-development/2015/jan/14/tanzania-illegal-logging-tree-species -extinction.

8. "Buying Farmland Abroad: Outsourcing's Third Wave," *The Economist*, May 21, 2009, accessed February 23, 2019, economist.com/node/13692889. See also Interview: "Neokolonialismus in Afrika: Großinvestoren verdrängen lokale Bauern," *Spiegel* online, July 29, 2009, accessed February 23, 2019, spiegel. de/wirtschaft/neokolonialismus-in-afrika-grossinvestoren-verdraengen -lokale-bauern-a-638435.html.

9. This issue is being examined by Chris Arsenault, thanks to a scholarship by the Pulitzer Center: pulitzercenter.org/education/meet-journalist-chris -arsenault, accessed March 10, 2019.

10. Blue Ventures, accessed February 23, 2019, blueventures.org/.

11. Campaign for Female Education or CAMFED, accessed February 23, 2019, camfed.org/.

Chapter 13

1. All the Footprint results in this chapter (and the entire book) are based on the 2019 edition of the National Footprint and Biocapacity Accounts. The results are freely available at data.footprintnetwork.org.

2. "World Population Prospects: The 2015 Revision," UN DESA, accessed February 28, 2019, un.org/en/development/desa/news/population/2015-report.html.

3. Greta Thunberg's original TedX talk is impressive: ted.com/speakers/greta _thunberg or visit her Twitter Page at https://twitter.com/GretaThunberg.

4. Searches with the Meltwater media tool identified the stories that contained "Global Footprint Network" as well as the audience size of each outlet. This allowed the network to determine the media impressions—or the number of people who could have seen the story on their media platforms.

5. Global Footprint Network evaluated Ecological Footprint reduction potentials for Earth Overshoot Day 2017 and 2018, using as references the 2017 and 2018 edition of the National Footprint and Biocapacity Accounts. These results are presented on the Earth Overshoot Day website at overshootday.org.

6. Examples for offset providers include myclimate.org, atmosfair.de, climate care.org, or greenseat.nl. WWF commissioned a study: Anja Kollmuss, Helge Zink, and Clifford Polycarp, "Making Sense of the Voluntary Carbon Market: A Comparison of Carbon Offset Standards," accessed March 11, 2019, global carbonproject.org/global/pdf/WWF_2008_A%20comparison%20of%20 C%20offset%20Standards.pdf. They argue for following the "Gold Standard" goldstandard.org/.

7. World Commission on Environment and Development, *Our Common Future* (frequently referred to as the Brundtland report after Gro Harlem Brundtland, Chair of the Commission).

8. Wilson, *Half-Earth* and natureneedshalf.org/.

9. Steve Pinker, *Enlightenment Now: The Case for Reason, Science, Humanism, and Progress* (New York: Viking, 2018).

10. "Japan Ecological Footprint Report 2012," WWF Japan: Tokyo, accessed on February 28, 2018, footprintnetwork.org/content/images/article_uploads /Japan_Ecological_Footprint_2012_Eng.pdf.

11. See Chapter 10.

12. See Herman Daly's illuminating remarks on the Plimsoll line in the context of his theory about the full and the empty world: Herman E. Daly, *Wirtschaft jenseits von Wachstum: Die Volkswirtschaftslehre nachhaltiger Entwicklung* (Salzburg: Verlag Anton Puste, 1999), 74 ff.

13. J. Rockström et al., "A Safe Operating Space for Humanity," *Nature* 461 (September 24, 2009): 472–475, and W. Steffen et al., "Planetary Boundaries: Guiding Human Development on a Changing Planet," *Science* 347, no. 6223 (February 13, 2015), doi: 10.1126/science.1259855.

14. Helmut Haberl, Karl-Heinz Erb, Fridolin Krausmann, "Global Human Appropriation of Net Primary Production (HANPP)" in *Encyclopedia of Earth*, ed. Cutler J. Cleveland (Washington, DC: Environmental Information Coalition, National Council for Science and the Environment), [first published in the *Encyclopedia of Earth* April 29, 2010] accessed February 28, 2019, editors.eol.org /eoearth/wiki/Global_human_appropriation_of_net_primary_production _(HANPP).

15. See for instance: William Macpherson, "Rwanda in Congo: Sixteen Years of Intervention," African Arguments, July 9, 2012, accessed February 24, 2019, africanarguments.org/2012/07/09/rwanda-in-congo-sixteen-years-of -intervention-by-william-macpherson, and Baobab, "Congo and Rwanda: Stop Messing Each Other Up," *The Economist*, July 3, 2012, accessed February 24, 2019, economist.com/baobab/2012/07/03/stop-messing-each-other-up.

16. Jevons, *The Coal Question*, 123–124.

17. "UAE Green Key Performance Indicators," UAE Ministry of Climate Change and Environment, no date, page 3, accessed February 25, 2019, moccae.gov.ae /assets/download/9c7ea0fa/uae-green-key-performance-indicators-pdf.aspx.

18. For details, visit masdar.ae, accessed February 28, 2019.

19. See Chapter 10.

Index

Page numbers in *italics* indicate figures.

About the Authors

MATHIS WACKERNAGEL, born in Basel, Switzerland, in 1962, is co-creator of the Ecological Footprint and president of Global Footprint Network, which was named one of the World's 100 Top NGOs by *The Global Journal*. Wackernagel has worked on sustainability with governments, corporations, and international NGOs on six continents, and has lectured at more than a hundred universities. He previously served as the director of the Sustainability Program at Redefining Progress in California and ran the Centro de Estudios para la Sustentabilidad at Anáhuac University in Mexico. Wackernagel has authored or contributed to over one hundred peer-reviewed papers, numerous articles and reports, and various books on sustainability. With dozens of international awards and accolades for his pioneering work, he has been identified as a leader who is driving the world's most significant problems to zero. He lives in Oakland, California. footprintnetwork.org.

BERT BEYERS, born in Mönchengladbach, Germany, in 1956, is a senior editor at the Norddeutscher Rundfunk in Hamburg. For several decades, questions of ecology and future have been his professional passion. He has published widely, including a book with Franz Josef Radermacher on survival in the 21st century entitled *Welt mit Zukunft: Die ökozoziale Perspektive*. He lives in Hamburg, Germany.

ABOUT NEW SOCIETY PUBLISHERS

New Society Publishers is an activist, solutions-oriented publisher focused on publishing books for a world of change. Our books offer tips, tools, and insights from leading experts in sustainable building, homesteading, climate change, environment, conscientious commerce, renewable energy, and more—positive solutions for troubled times.

We're proud to hold to the highest environmental and social standards of any publisher in North America. This is why some of our books might cost a little more. We think it's worth it!

- We print all our books in North America, never overseas
- All our books are printed on 100% **post-consumer recycled paper**, processed chlorine-free, with low-VOC vegetable-based inks (since 2002)
- Our corporate structure is an innovative employee shareholder agreement, so we're one-third employee-owned (since 2015)
- We're carbon-neutral (since 2006)
- We're certified as a B Corporation (since 2016)

At New Society Publishers, we care deeply about *what* we publish—but also about *how* we do business.

Download our catalog at https://newsociety.com/Our-Catalog or for a printed copy please email info@newsocietypub.com or call 1-800-567-6772 ext 111.

WHAT IS THE FOOTPRINT OF THIS BOOK?

- If this book had been printed on virgin, coated paper, its Ecological Footprint would occupy over 20 global square meters (~200 square feet) for one year.
- Because this book is printed on 100 % post-consumer recycled, non-coated paper, its Ecological Footprint amounts to just one quarter, or about 5 global square meters (~50 square feet) for one full year, assuming conventional processing energy.
- If all the processing energy came from wind & solar power, the Ecological Footprint would be one fifth, or 1 global square meter (~10 square feet) for one year.
- Five square meters, or about 50 square feet, is equal in area to all the pages of this book combined. This means that the book's Ecological Footprint occupies the same space for one year, as the sum of its spread-out book pages.

Data for Ecological Footprint assessment stem from the Environmental Paper Network's papercalculator.org.